Face Value

Face Value

Physiognomical Thought and the Legible Body in Marivaux, Lavater, Balzac, Gautier, and Zola

Christopher Rivers

THE UNIVERSITY OF WISCONSIN PRESS

The University of Wisconsin Press
114 North Murray Street
Madison, Wisconsin 53715

3 Henrietta Street
London WC2E 8LU, England

Library of Congress Cataloging-in-Publication Data
Rivers, Christopher.
    Face value: physiognomical thought and the legible body in Marivaux,
  Lavater, Balzac, Gautier, and Zola / Christopher Rivers.
    288 p.   cm.
    Includes bibliographical references and index.
    ISBN 0-299-14390-2   ISBN 0-299-14394-5 (pbk.):
    1. French literature—19th century—History and criticism.
  2. Marivaux, Pierre Carlet de Chamblain de, 1688–1763
  —Knowledge—Physiognomy.
  3. Lavater, Johann Caspar, 1741–1801—Knowledge—Physiognomy.
  4. Physiognomy in literature. 5. Face in literature.
  I. Title.
  PQ283.R58   1995
  840.9'007—dc20   94-14586

For Christopher Miller

Contents

Illustrations

Acknowledgments

This book would not have been possible without the generous help of many individuals and institutions. *Face Value* began as a dissertation in the French department at Yale University, and benefited at several stages from careful and encouraging readings by Ora Avni and Charles Porter. Peter Brooks was a thoughtful, consistent, and insightful advisor; my debt to his work will become apparent to any reader of this book. I was able to complete a first version of the manuscript because of a generous fellowship from the Mrs. Gilles Whiting Foundation. I am also grateful to the courteous and helpful staff of the Beinecke Rare Book and Manuscript Library at Yale.

Over the past few years, I have been fortunate enough to be able to meet a number of the scholars who have written on Lavater, and I would like to thank each of them for his/her inspiration: John Graham, Barbara Maria Stafford, Judith Wechsler, and especially Graeme Tytler. Special thanks go to Ellis Shookman for organizing a wonderful conference at Dartmouth in honor of the 250th anniversary of Lavater's birth, and thus providing Lavater scholars with the opportunity to get to know each other.

I would also like to thank various colleagues and friends who have offered either material, intellectual, or moral support in the arduous task of seeing a book through to publication: colleagues in the French department at Smith College, in whose company I spent a most pleasant year; the staff of Williston Memorial Library of Mount Holyoke College; my colleagues Jena Gaines and Kavita Khory; members of the Pew faculty seminar at Mount Holyoke on representations of the body, especially our leader Karen Remmler; and my colleague Linda Jones Jenkins and friends Leyla Ezdinli and Jennifer Wicke, each of whom has always had the great generosity to listen patiently to a laundry list of complaints and anxieties. Special thanks must go to all my colleagues, past and present, in the French department at Mount Holyoke College—Alexandre Albert-Galtier, David Ellison, Jack Erwin, Elissa Gelfand, Samba Gadgigo, Catherine LeGouis, Jacques-Henri Périvier, Margaret Switten, and Nicole Vaget—for all of their support; to Claire Benoit; and to all the wonderful students whom it has been my pleasure to teach at Mount Holyoke.

Everyone at the University of Wisconsin Press has been efficient, encouraging, and courteous, and for that I am profoundly grateful. Many thanks are due to Barbara Hanrahan, Mary Elizabeth Braun, Margaret Walsh, Colleen Heinkel, Raphael Kadushin, and especially Allen Fitchen. Susan Tarcov was an excellent copy editor, and I am grateful for her skill and hard work. I would also like to thank Ross Chambers and Mary Lydon, who were generous and perceptive readers of the manuscript. My work was inarguably improved thanks to their useful comments.

On a more personal note, I would like to thank my parents, Charles Ford Rivers, Jr., and Elizabeth Dufour Rivers, who have always been supportive of my academic endeavors; and my grandparents, Charles Ford Rivers, Sr., Ella Moseley Rivers, Alfred Emmanuel Dufour, and Louise Baker Dufour, each of whom has been an inspiration in some way.

Last, and most important, I would like to express my limitless gratitude to Christopher Miller, who has supported me completely and in every way imaginable, from the earliest glimmer of my somewhat offbeat idea for a dissertation to the final stages of the construction of this book. Without him, the work would have been neither possible nor worthwhile.

Face Value

Introduction

In French, the word *physionomie* is defined as either "l'ensemble des traits, l'aspect du visage" or, in a second definition, "aspect particulier propre à une chose, à un objet." The word *physiognomonie*, by contrast, is defined as "science qui a pour objet la connaissance du caractère d'une personne d'après la physionomie." Unfortunately for the purposes of the following study, English does not make such a distinction, "physiognomy" serving to refer both to physical appearance and to the pseudoscience of divining character from physical traits. In the chapters which follow, I treat both referents; physiognomy is pertinent to my study as a general mode of perception and representation reflected in pseudoscientific and especially fictional narrative discourse, rather than as a specific intellectual and cultural influence. As an interpreter of literature, I emphasize the rhetorical and narrative, rather than historical and cultural, expressions of what I refer to as "physiognomical thought," a term which I define somewhat broadly for the purposes of this study. Generally, I consider physiognomical any supposed correlation between the physical and metaphysical realms, particularly when the physical is pressed into service as a semiotic index to the metaphysical. "Physiognomical thought" therefore always functions as a semiotic system, in which the body is read as a signifier and character, essence, sociopolitical status, or gender as its signified. It is this act of semiotic interpretation, and more particularly its ramifications for our understanding of certain narrative texts, that will interest me here.

My work is somewhat related to, but not directly inspired by, the recent wave of scholarly work on representations of the body. Anyone having worked recently in the fields of literary or cultural studies, of art or intellectual history, or indeed in any one of many different disciplines is familiar with the ubiquity of "the body" as an intellectual topos. Just a few months ago, I was in the Amherst College bookstore looking for a book that was curiously absent from both the women's studies and literary criticism shelves; upon asking, I was informed by the clerk that I would find it in the "body crit" section. Indeed, to my amazement, there were (and still are, I presume) several shelves of books which compose a full-fledged "body crit" section—a convincing testimony to both the scope and the popularity of the recent trend. As I will discuss below in greater detail, my work is at once sympathetic with, and inarguably distinct from, most of the work classified as "body crit." Indeed, as the only full-blown rhetorical reading to date of the eighteenth-century Swiss theoretician Lavater (the most widely known modern physiognomist), it distinguishes itself clearly even from other studies of physiognomy.

Face Value is based on some elementary premises which differ from those emphasized in many works of "body crit": that the body is central to the functioning of many works of narrative fiction, that these works are full of instances of body reading, and that the body is as legible and as illegible a text as any other. Most important, I attempt to demonstrate that the most effective and interesting way of discussing these phenomena is through a close reading of individual literary works as more or less autonomous textual constructs. Furthermore, I believe that representations of the body necessarily mean different things in the contexts created by different works. In writing *Face Value*, I chose to take an approach that would have been unthinkable before the literary-critical "revolution" of the late 1960s; that is to say, I examine individual themes within individual works more or less exclusively on the terms of those works and, for the most part, not as they may or may not be said to relate either to other texts or to the historical "reality" of the period in which they were written. I try to discuss explicitly the esthetic implications and ramifications for specific works of the theme of the "legible body" (I share the term with Michael Shortland). In other words, I take as an important component of my work within each chapter the elucidation of the significance of the body to our understanding of the novel in question as a linguistic, rhetorical, and narrative construct. If the answers to the questions I pose vary widely from novel to novel, the fundamental concerns remain constant and intertwined.

Another of my premises is that the *degré zéro* of even the most complex rhetorical and narratological analyses is inevitably a *thematic* reading: the legible body is indeed a theme, and a work such as mine takes as its most obvious tool certain thematic recurrences and variations within a chosen set of texts. I would argue, however, that the theme of the legible body is unlike most others and therefore worthy of scrutiny: it is privileged in that it transcends the realm of raw theme by the important questions it raises about larger literary issues.

Face Value will suggest that a reconsideration (and rehabilitation) of the role of theme in literary analysis might thus be a valid undertaking for critics: by this, I do not mean a return to earlier, unreflective models of thematic criticism, wherein themes are catalogued with little attention to their significance. I intend rather to undertake readings which will use theme as a tool for complex textual analysis, which will read theme as a potential figure of rhetorical and narratological interpretation.[1] If, as Rousset so long ago demonstrated in *Forme et signification*, form is inextricably linked to signification and therefore must be considered in any reading of a literary text, I would argue that the same may be true of theme, which, as the referential system of a literary work, constitutes the raw material of signification. Theme, as Paul de Man and others taught us, is really nothing but "referential aberration." It is, however, an aberration that is constitutive of novels—particularly those I have chosen to study here. A theme such as that of the legible body can therefore serve as a gateway to the most significant questions a novel poses. Thus, while I would certainly not argue that all themes or all works of literature would lend themselves equally well to such a thematic approach, I would maintain that certain themes in certain works provide invaluable and unquestionable means of access to questions of transcendant importance to any literary-critical treatment. A genuine linking of theme and signification was the criterion I used in choosing the works for *Face Value*, and my work aspires to be a sort of "enlightened" thematic criticism.

This is indeed the very crux of my project, which attempts in each case to take a thematic reading and link it, through the theme itself, to significant questions about both the functioning and the significance of the narrative work in which it is situated and which, in some cases at least, it can be said to constitute. Physiognomical thought in literature, which thematizes acts of reading, interpretation, and knowledge, obviously lends itself particularly well to such an exploitation on the part of a literary critic interested in rhetorical reading, and therein lies its interest.

As will be obvious from the chapters that follow, I define narrative primarily in its function as a means of ordering, of creating intelligibility; I consider all narrative discourse to be the creation of—or, at least, a reflection on—systems of intelligibility, coherence, and, by extension, knowledge.[2] As attempts to create coherent (and thus reassuring) explications of human life, all works of more or less traditional narrative rely on the assumption that the narrator and/or author knows whereof he speaks, on his ultimate authority. (For reasons that will become obvious below, my use of gendered pronouns in the context of this discussion is not arbitrary.) The same is true of all physiognomical discourse per se, and indeed of all physiognomical thought, in which the physiognomist/author implicitly or explicitly demonstrates his skill at making bodies signify and thus at explicating human life to a necessarily rather passive and often mystified reader.

In the case of fictional narratives, physiognomical themes serve as a sort of metanarrative of explication. Some characters' bodies reveal in some way the truth about them, and it is this truth which governs the narrative.[3] In each work I have chosen to study, the central question around which the plot is ordered is one of identity of some kind: who is Marianne, the beautiful foundling? who is the girl with the golden eyes? who will be the better husband, the effete and elderly aristocrat or the hairy, muscular Bonapartist? is the beautiful young cavalier and duellist a man or a woman in men's clothing? where does the demimondaine come from and what lies inside her pink and gold corpulence?

Indeed, these novels might be said to constitute a strange sort of physiognomical subgenre not unlike the *roman policier*, in which the question which fuels the narrative is not "whodunit?" but "what does this body mean?" One of the most basic threads which tie my various readings together is in fact the importance in all of these texts of the process of deciphering corporeal and physiological signs in an attempt to gain access to a higher comprehension of human existence— or at least greater knowledge of an individual. Each of the authors I have chosen alternately asserts, thematizes, and questions the possibility of the body as signifier. In each case, the reader is ultimately provided with the answer; in none is the answer explicated. As I shall discuss again in my conclusion, we are informed by the texts and the stories they relate that bodies do signify, but are never shown how the pivotal act of interpretation might take place. Significantly, if surprisingly, this is as true for Lavater's physiognomical treatise as it is for the novels in question.

As the corporeal is purportedly explained through its supposed relation to the metaphysical, it becomes a privileged site for what has been called, in reference to Balzac, "hypersignification."[4] The will to create meaning, order, and coherence through the complementary processes of signification and narration leads authors such as Balzac and Zola to ascribe an inordinate power of signification to the human body. Echoing Lavater and other theorists of physiognomy, either consciously or (more likely) by coincidence, these authors seem to take as their most fundamental agenda the denial of the arbitrary nature of human existence. Fictional narrative in nineteenth-century France, exemplified by Balzac and Zola, almost invariably entails a reinforcement of the notion that there exist answers to the questions and contradictions apparent in human life. These works make obvious the fact that narrative is by definition a discourse of explanation and illumination. Again, the fundamental importance of the thematics of physiognomical concerns in these works lies not, therefore, on the level of theme but rather in their implications about the nature of narrative discourse itself. As we shall see in the cases of Lavater, Marivaux, Balzac, and Zola, various examples of narrative discourse share a common agenda: that of the "study of man." Marivaux's disabused spectator, Lavater's "genius," Balzac's observer of the human heart, and Zola's scientist all attempt to see the invisible, know the unknowable, and narrate that which lies beyond the limits of narration.

If we as readers have access to the visible and the knowable through the narrative itself, we do not have access to the ruling metanarrative of corporeal legibility. In other words, the text lets us know that body reading is at issue, and ultimately constitutes itself an example of physiognomical reading, without explaining how such readings take place and what rhetorical, conceptual, or scientific logic informs the process. The answers to those questions remain the exclusive domain of the physiognomical genius—the narrator and/or author of the text. A certain passivity on our part vis-à-vis the act of interpretation of the text is thus assured. The logic of the text is ordered by systems of interpretation not explicated within it. In these texts, accurate explanations of why the central character is the way she is are impossible for the reader without physiognomical elucidation, which is never present as such in the text. The text remains clearly ordered, but by an order exterior to itself and thus beyond our grasp as readers. We may observe the unfolding of physiognomical tales, but necessarily lacking the (physiognomical) knowledge which determines them, we cannot fully participate in the production of meaning they represent. The texts studied in *Face Value* are "writerly" texts par excellence, in which the very themes of the works assure that the author will maintain his authority over their interpretation. We as readers have more in common, then, with the characters of the novel, who invariably remain outside of full knowledge, than with the omniscient author or implied narrator.

However, in sharp contrast to all the phenomena discussed above, Gautier, through the medium of his antigenius d'Albert, in *Mademoiselle de Maupin*, puts into question "the study of man." Unlike the texts of his predecessor Marivaux, his contemporary Balzac, or his successor Zola, which implicitly or explicitly allude to the answers to the corporeal questions they pose, thereby paralyzing the reader in the presence of superior but implicit knowledge, Gautier's text itself

ultimately participates in the most profound questions facing the characters in his novel. For much of the novel we know the answer to what
appears to be the central corporeal question of his novel: is "Théodore" a man or a woman? However, the conclusion of the novel turns
our certainty of what gender means on its head and thereby poses
a much more profound and perhaps unanswerable question. In so
doing, Gautier ultimately problematizes the relation between knowledge and narrative and emphasizes that which Balzac and Zola seek
to obliterate: the ambiguous, the arbitrary, the inexplicable.

With the body itself pressed into service as a figure of interpretation in all of these novels, various systems are established whereby all
ambiguity can at least potentially be eliminated. This is of course not
to say that the characters within each fictional text have access to the
system; indeed, physiognomical reading is invariably problematized
at the level of plot and it is these very problems that serve to drive
the narrative. What is most important to note again, however, is that
the author is always in possession of the explanatory metanarrative,
and thus, at least on one level, in complete authority. The novels I
have studied (with the notably subversive exception of *Mademoiselle de
Maupin*) are works of literature to which there do exist predetermined
answers, imposed interpretations, and would-be "scientific" (and, one
is forced to assume, extradiegetic) explanations of the events of the
narrative.

Nonetheless, in keeping with physiognomical tradition as exemplified by Lavater, how one arrives at these answers is often unclear at
best. One must know how to read bodies properly in order to arrive
at the knowledge always already possessed by the author and thus
to participate fully in the text; unlike much of the common sense or
cultural literacy required for an intelligent reading of most fictional
texts, however, the physiognomical skill and knowledge required for
an active understanding of these texts are not attainable within the text
itself and are beyond the scope of most readers. Since physiognomical
practice is obscure and disjointed even in treatises on the subject such
as Lavater's, no cogent explication of it even seems possible. In both
pseudoscientific and fictional discourse, physiognomical skill, it is implied, is the fruit of some unspecified form of innate "genius." Within
these narratives, the secret of physiognomy can never be wholly un-
veiled, for, like the tricks of a faith healer, it will not withstand the
light of day. Indeed, one might well argue that the very notion of
physiognomical knowledge is a fiction.

It should thus come as no surprise that the "reading" characters
in these novels, those who (knowingly or not) would be physiog-

nomists (Valville and Mme. de Miran in *Marianne,* Henri de Marsay in *La Fille aux yeux d'or,* Mlle. de Cormon in *La Vieille Fille,* d'Albert in *Mademoiselle de Maupin,* all the men and especially Muffat in *Nana*), are invariably misguided in their assumptions about what a particular body means. If all narrative is "about" intelligibility, these particular narratives foreground that fact by positioning their characters in various ways vis-à-vis the potentially legible and intelligible truth of the body in question. Without access to a higher knowledge of the corporeal truth of the character whose essential identity is in question (such as that possessed by the author), they are doomed to misreading. As readers of such texts, we have a choice: either to resign ourselves to a passive acceptance of the superior knowledge the text represents, or to perform a skeptical critique of the notion of physiognomy itself. As the following chapters will make clear, the purpose of this study is to take the latter approach to a number of texts in which this problem of reception is particularly acute.

Physiognomical thought is by definition essentialist and deterministic, the body always serving as an index to some fundamental, innate truth which orders and explains an individual. The potential sociopolitical ramifications of such a worldview are obvious, and provide another link among the works in this study. While theories which attempt to establish a deterministic link between corporeal and moral or intellectual traits have been often used as tools of racist ideology, the texts I chose for this study show the power of such ideas as tools of sexism as well. Such a reading is of course informed by late-twentieth-century social and political beliefs and is not in any way an attempt to ascribe anachronistic intentions to the authors in question. For readers even marginally conversant in the various feminist approaches to literature articulated over the last few decades, however, the implications of these texts with respect to gender are vast and compelling.

At the risk of oversimplifying, can it not be said that all sexism is metaphysical determinism based on sex, and, as the most basic definition of sex is corporeal, that sexism is therefore always to some extent the creation of an imaginary correspondence between physical and metaphysical domains? What is sexism if not the ascribing of common moral, characterological, and intellectual qualities to a group of people based solely on their common anatomical features? Does this definition not echo the notion of physiognomical thought itself, and would it not find its most appropriate expression through the "explanatory" medium of narrative? If we agree with the above, it should come as no

surprise that texts such as those studied here, in which physiognomical concerns are central, should turn out to be texts which are also,
to our late-twentieth-century sensibilities, inherently and undeniably
misogynist.

It was not my intention to write a work of feminist criticism, and
there are indeed no explicitly feminist theoretical underpinnings to
my study; however, the role of the female body and of physiognomy
in the novels leads me directly, and inevitably, in each instance, to reflection on compelling questions of gender. Each of the fictional texts
I have analyzed takes as its title, or part of its title, the name of a
woman (*La Vie de Marianne, Nana, Mademoiselle de Maupin*), a generic
classification of woman (*La Vieille Fille*), or an appelation referring to
an individual physical trait of a woman (*La Fille aux yeux d'or*). All are
texts which have a woman, and indeed a woman's body, as the central
figure, and as an object of analysis. Not only are women's bodies traditionally more objectified as mere material entities than men's, they
are more often read as objects of (male) interpretation as well. In the
fictional narratives I study, women's bodies often act as obstacles to
this reading by provoking desire in men and, in some cases, admiration and/or jealousy in other women. The emotions provoked by
the female body as spectacle generally direct the plot of the novels,
with the notable exception being Balzac's *vieille fille* whose own desire
clouds her judgment and helps to render her grotesque.

From the perhaps well-intended but misguided "glorification du
principe féminin" in *La Vie de Marianne* to the pathologically misogynistic myth of the *femme fatale* in *Nana*, the texts in this study demonstrate several of the many ways in which women's bodies have been
interpreted as powerful, and often dangerous, objects, and the ways
in which the novel can become a forum for such myths. As I attempt
to show in my chapter on the history of physiognomy, women have
always been considered separate, enigmatic, problematic, and sometimes distasteful objects of physiognomical thought.[5] Barbara Stafford
has in fact made the astute link between gender, the Enlightenment
mania for logic and order, and physiognomical thought: "Physiognomics was 'Neoclassical' and male in its linguistic and single-minded
will to impose sequence and logic on experiential confusion."[6] If most
theoreticians of physiognomy itself, such as Lavater, tend to avoid
analysis of female bodies, physiognomical novelists by contrast seem
interested in little else. The novels seem to take as their agenda the
representation, interpretation, and explication of the female body. The
texts I have chosen exemplify this better, in their various ways, than
any others in the French canon, and each poses some rather pointed

questions about the legible body and its relation to sex roles and the meaning of gender.

This leads me to a remark about methodology: at the risk of appearing to create a paradox, I would affirm (as, of course, have many others before me) that it is seemingly ahistorical close reading methods of textual analysis which are particularly effective in teasing out the sometimes subtle and implicit ideological connotations, as well as esthetic and rhetorical workings, of a work of either fiction or nonfiction. It is only through a literary reading of both the novels and the "nonliterary" texts in the study that I am able to discuss the link between physiognomy and misogyny they represent.

Thus, I have chosen the texts for this study because they pose similar ideological, as well as "purely" literary, questions; I would indeed argue that these are texts in which the two registers are inextricably linked. They are not merely works in which physiognomical themes are *mis en scène*, as this could be said of almost any work of narrative fiction written in the eighteenth or nineteenth century. They are works in which these themes most explicitly control and constitute the narrative, representing significant ramifications for both esthetics and ideology, and thereby transcending the level of mere theme. In these works, theme becomes so explicit and so dominant as to be inextricable from the rhetorical function of the text.

Each of the novels I have chosen, written by a man, is unquestionably "about" the female body and "of" the female body: the theme of body reading is the motivating force on the levels of both *histoire* (plot) and *récit* (narrative). It is the centrality of the body to both the story and its telling which distinguishes these novels from others of the period in which, I would argue, the body remains a mere theme—*Madame Bovary*, for example. If, as I have asserted above, narrative in general implies a will to explicate, in the novels I have chosen to study, this general impulse is reinforced by the specific notion that the ruling explanatory metanarrative of corporeal legibility is both physiognomical and (invariably) misogynist. It is this phenomenon which links the texts in *Face Value*, and renders them subtly distinct from others in the eighteenth- and nineteenth-century French canon.

In the interest of elucidation of the methodological premises of my work, it would perhaps be useful for me to sketch out, albeit briefly, the scholarly works in relation to which my study might be defined. The examples I discuss below will serve to demonstrate that my work has little in common with intellectual history, a bit more in common with literary history, but is primarily a work of literary criticism.

Many interesting attempts, from various perspectives, have been made in recent years to create an intellectual history of systems of representing and interpreting the body, both in the United States and in France. Among the most notable treatments of the period which interests me are Barbara Stafford's previously cited *Body Criticism,* Courtine and Haroche's *Histoire du visage,* Daniel Roche's *La Culture des apparences,* and Philippe Perrot's *Le Corps féminin: le travail des apparences.*[7]

Stafford is interested in (among other things) physiognomy per se; as a sort of nexus of artistic and medical representation, physiognomy is indeed the ideal object of analysis for her study. Her commentary on Lavater is extensive, intelligent, and compelling. Nonetheless, there is considerably less overlap between her work and my own than one might imagine at first glance. The explicit agenda of Stafford's impressively erudite *Body Criticism* is to contribute to further understanding of the "language(s)" of the visual. Stafford argues against the textualization of art in much art criticism and advocates that images be "read" and understood on their own terms *as visual objects. Body Criticism* represents a choice on Stafford's part to examine the visual, rather than verbal, metaphors of the body in eighteenth-century art and science; in keeping with her agenda, she performs a provocative and original analysis of Lavater's work through its pictorial elements. By contrast, my work explicitly focuses on literary and rhetorical questions, on issues directly and inextricably linked to the written text and to textuality itself.

Significantly, while maintaining the emphasis on images and visual analysis that is the foundation of her project, Stafford herself frequently acknowledges the link between textual analysis and some of the issues she studies: "The activity of searching inquiry linked medical diagnostics to textual criticism and cerebral expertise."[8] As these are the kind of links which I consider fundamental to my own understanding of physiognomical thought, this passage suggests to me that Stafford and I share at least a basic understanding about corporeal metaphors. One might contrast our approaches to Lavater by saying quite simply (although it is in fact considerably less simple than this) that Stafford looks at the pictures and I at the words. In short, Stafford's work and mine are diametrically opposed in approach and agenda while essentially sympathetic in philosophy.

Another recent intellectual history of the legible body is Courtine and Haroche's *L'Histoire du visage,* which examines the paradox established by what the authors see as two contradictory discourses about the body, one shaped by the wide range of corporeal expression (what

I would call the "legible body") and the other by the importance of controlling the language of the body in the interest of "civility." Courtine and Haroche explore physiognomical thought from the sixteenth through the nineteenth century and continually contrast its premise of legibility with contemporary imperatives concerning the necessity of masks, of remaining impassive and illegible in society. Their topic is extremely well conceived, and makes for a fascinating study of two conflicting historical discourses. Its relation to my own interest, the role of the legible body in fictional narrative discourse as such, is thus an interesting but indirect one.

Roche's book, as its subtitle ("une histoire du vêtement . . .") makes clear, takes as its object of study clothes in the seventeenth and eighteenth centuries in France; it is a fascinating social and anthropological history of the production, consumption, and semiotics of dress in the period. It is therefore related in an obvious yet secondary way to some of the questions which interest me. Chapter 14, "Les Vêtements de roman," is notable in that Roche discusses there literary representation of clothes. As is appropriate for the kind of historical survey he is writing, however, his commentary on clothes in literature remains for the most part rather general (with exceptions being his discussions of Casanova's *Mémoires* and Rousseau's *Emile*). He does not engage in textual analysis per se and links his discussion of the semiotics of clothes to larger social and historical, rather than esthetic and rhetorical, issues.

Roche's notion of the central role of clothes and other corporeal themes to French literature of the eighteenth century is nonetheless very much like my own. For example, he says: "L'esthétique du roman des Lumières est fort heureusement pour nous inséparable d'une conception du corporel, même si nombreux sont les récits silencieux sur les détails matériels vestimentaires."[9] As with Stafford, the contrast between a study such as Roche's and mine is less one of philosophy than of methodology. Thus, I would concur entirely with the passage I have cited, but would widen its scope to include all literary description of corporeal signs in the period to which he alludes, not merely that representing clothing. My work might thereby seem to enlarge the field of Roche's semiotics by extending from clothing to the body and physical appearance in general; on the other hand, it limits the objects of study to literary representations of the body and examines them on their own terms as particular exemplars of fictional narrative discourse. My work is thus at once much wider and much narrower than *La Culture des apparences*.

A similar analogy is true of the relation my work might be said to have with Philippe Perrot's *Le Corps féminin*. As a socio-psycho-historian, Perrot is concerned with the many different perceptions of the female body, both "real" and represented, in the eighteenth and nineteenth centuries, and we share a belief that gender must be central to any discussion of corporeal discourse. While Perrot's allusions to works of literature are frequent and intelligent, he does not discuss literary representation as such. Given his goal of uncovering the psychological, historical, and cultural import of representations and perceptions of the body, he understandably chooses not to focus, as would a literary critic, on the esthetic and rhetorical ramifications of such representations.

In each case described above, my differences with the work are ones of methodology and discipline rather than philosophical content itself. In spite of all the unquestionably invaluable possibilities for blurring the lines among the disciplines that recent scholarship has demonstrated, comparisons between works of intellectual history such as those I have cited and works of textual analysis like my own will always be of limited validity. I find the work done by historians on topics analogous to my own fascinating and impressive; however, my goals, preoccupations, and methods are distinctly different from theirs. As literary criticism, my work seeks to explicate and analyze works of literature as autonomous linguistic and esthetic constructs and not as historical or anthropological documents. This fundamental contrast in disciplinary agenda must be understood in order to appreciate the intent of my work here: *Face Value* seeks to study physiognomical thought, as expressed through the theme of the legible body, in carefully chosen works of narrative literature, because of the richly suggestive questions it poses for the study of literature per se. These questions and the answers to them vary greatly among the heterogeneous group of texts I have chosen; as I have discussed above, however, my discussion of each work treats fundamentally literary issues: those of representation, of rhetoric, and of narrative.

Closer, then, to my own work is the only comprehensive literary-historical study of physiognomical themes in fiction: Graeme Tytler's invaluable *Physiognomy in the European Novel: Faces and Fortunes* is the only book-length study of the relation between physiognomy and literature to date.[10] Indeed, Tytler's intelligent and insightful work greatly influenced my own and must serve, along with John Graham's groundbreaking studies of Lavater, as a primary reference work to any scholar interested in the topic. Tytler's book is, however, at bottom a work of literary *history*: an impressively erudite, seemingly exhaustive survey of physiognomical themes in English, French, and German literature, with a special emphasis on the nineteenth century, but treating works from the medieval period to the twentieth century. With a few exceptions, the extremely ambitious scope of his project prevents Tytler from spending more than a paragraph or two on any individual literary work. By contrast, the methodology I employ is that of close reading of a few works, chosen for their interest as exemplars of textual phenomena. Fortunately for scholars interested in the question of physiognomy and literature, Tytler has already written the definitive literary-historical view, thus affording others such as myself the luxury to undertake more specialized analyses.

Another work of relatively recent criticism with which *Face Value* may be both compared and contrasted is Helena Michie's very provocative and intelligent *The Flesh Made Word*. Most fundamentally, Michie and I share an interest in the misogyny inherent in much of the canon of literary representation of women's bodies. We also share, very roughly speaking, certain elements of a poststructuralist literary-critical perspective and a commitment to rhetorical reading. Our work is thus undeniably compatible.

However, if Michie's work seems somewhat closer to mine in its sensibilities, ideological preoccupations, and methodological influences than the others cited here, it—like the others—also presents some very important points of contrast. Although our work is perhaps influenced by some of the same intellectual currents, Michie includes as an important part of her agenda a direct treatment of poststructuralist and feminist thought, while I have opted to undertake a literary-critical, rather than explicitly theoretical, approach to my subject.

Furthermore, in contrast to the corpus of my study, Michie's work concerns itself largely with Victorian novels and conduct manuals, Pre-Raphaelite paintings, and contemporary American feminist criticism and poetry. In addition, there are important structural contrasts between my work and Michie's: most of her chapters and subchapters are thematically determined: that is to say, she takes a theme, image, or trope and discusses its various literary and other manifestations and implications, rather than taking a specific author or text as her primary unit of study. Within the selected historical and thematic limits of her study, then, Michie's work functions as something more like a survey than does *Face Value*.

Finally, there does exist a work somewhat more analogous to my own in terms of scope, discipline, object(s) of study, and (very roughly speaking) methodology than the others I have discussed: Roger Kempf's important *Sur le corps romanesque* (1968). While Kempf does not develop his discussion in the same way I do, the seeds of some of the notions which are most central to my work are clearly present in his book. He states, for example, in his introduction:

> De cette copulation du corps avec le livre va naître le corps du livre, le corps romanesque . . . L'on devine à ces exemples [the works he has chosen to study] que le corps est plus qu'un 'thème,' qu'il est constitutif d'un langage original et qu'il commande ma captation et ma captivité, mon entrée et ma présence dans le livre.[11]

My own statement of the notion Kempf puts forth above about the role of the body as literary theme would be considerably simpler, but would express roughly the same thought about its centrality and importance in certain works of narrative literature. Although the comments above are not developed per se in *Sur le corps romanesque*, they clearly reflect a fundamental conceptual similarity between my own work and Kempf's. Furthermore, and perhaps even more important, we each choose to proceed with our project on a text-by-text basis, analyzing chosen individual works on their own rhetorical and esthetic terms and avoiding overarching generalizations.

That said, I should specify that Kempf's literary analyses and my own are vastly different in actual content: first, in spite of its title, Kempf's book does not take solely corporeal features as its objects of study. He also treats such diverse and nonphysical signifiers as language, names, and manners. Second, and much more important, Kempf's analyses, in keeping with the style of the period in which they were written, are considerably more abstract and "philosophical" than my own, which tend to examine the works in question in their textual specificity. For example, although Kempf alludes both directly and indirectly on many occasions to the links between language, narrative, and the body, nowhere is this discussion explicitly developed or synthesized either on the level of an individual work or with respect to the chosen texts as a group. By contrast, my analyses aim to be direct, accessible, and explicit in their attempts to articulate how and why the body is more than a mere theme in the works I have chosen, and what its specific ramifications are for these individual works.

Although physiognomy is, and will probably continue to be, an extremely rich topic for intellectual and literary historians, it is no doubt clear at this point that my goal in *Face Value* will be to articulate the importance of physiognomical themes in literary works not as potential footnotes to intellectual or literary history, but rather as figures of literary interpretation.

I wholeheartedly support recent scholarly enthusiasm for interdisciplinarity, and recognize that it is indeed this phenomenon which implicitly authorizes a topic such as mine, allowing me to analyze both literary and nonliterary texts in a single study. However, I would also maintain that at least in some cases there is much to be gained from what might appear to be a single-minded methodology. Literary-critical methods such as those I attempt to employ can afford greater insight into not only literary texts, but also texts such as Lavater's, which are at least as rich when read as semiotic and rhetorical constructs as when considered "purely" historical documents. The historical import of a text like Lavater's is inevitably related to, and explicable through, its rhetoric.

We cannot possibly say with any certainty what effect these texts may or may not have had on behavior, on history, or on extratextual reality, and it is debatable whether or not an attempt to do so would be an interesting endeavor. We can however discuss how they function as texts in both the ideological and esthetic registers, and thereby better educate ourselves about the nexus of theme, rhetoric, narrative, and ideology that literary texts constitute.

Whether readers will find this book shelved in the "body crit," "lit crit," or any other yet to be devised section of their bookstores, it is my hope that the following readings will themselves justify my methodological choices and that my work will make a useful contribution to the ever-growing corpus of scholarly work on the body in literature.

Chapter One

From Analogy to Causality: The History of Physiognomy before 1700

The history of physiognomy is long and complicated. From the earliest periods of recorded history to today's studies of "body language" and morphopsychology, the notion of reading the human body seems to be of eternal fascination. Although the approaches to this question, both theoretical and practical, have varied greatly over the centuries, the fundamental concern, that of making the body signify, has remained constant. Before embarking on detailed discussions of specific treatments of this question, it would no doubt be useful to sketch here at least a brief outline of the history of physiognomical thought.

Mesopotamia

As early as the paleobabylonian period, handbooks for reading the body existed.[1] As Jean Bottéro tells us, physiognomy in the period was largely connected to practices of divination and prognostication, and indeed the human body was the primary object of study for Mesopotamian diviners. Much of the physiognomy of the time centered on such "abnormalities" as birthmarks and spots on the face. It is important, however, to note that even in this earliest period of physiognomical speculation, there is some division between divination that takes the human body itself as the object of interpretation and that which studies human behavior (including more or less corporeal categories as dress and gait, which eerily prefigure Balzacian preoccupations). The distinction between the body as immutable, fixed sign system and as (more or less) voluntary expression of consciousness is one which will continue to be important throughout the history of physiognomical thought, as we shall see in subsequent chapters.[2]

From this period until as late as the seventeenth century, physiognomy, as a system of corporeal semiotics, was linked not only to divination and the occult, but also to medicine. In spite of a seemingly contradictory and pervasive supernatural bent, Mesopotamian

18

physiognomy, as an exercise in deductive thought, represents an early step in the history of scientific method, as Bottéro has convincingly argued.[3] However, later in the period, the distinction between divinatory and medical physiognomy became increasingly clear as divinatory practice continued to entail, at bottom, the creation of a fictional narrative, while medical semiotics became rather more concrete and circumspect. A telling example of this divergence is given by Bottéro: a "livid" face might be read by diviners as the sign of a death by water, a death resulting from an oath not honored, or even as the sign of a long life; by contrast, medical semioticians would read it, somewhat more logically, as simply the sign of impending death.[4] It is to be noted, however, that the tension between supernatural and natural schools of physiognomy was not definitively resolved (in spite of the triumph of physiological models in the seventeenth century) until at least the nineteenth century, as such conflicting philosophies are still very much present in Lavater's late-eighteenth-century work on physiognomy. In fact, even as Bottéro outlines the distinction between the two physiognomical approaches, he maintains that divination *as science* is perhaps the most important contribution of the Mesopotamians to intellectual history.[5]

The specifics of the relationship between early science and the occult is less interesting to a student of physiognomy than the fact that the two "disciplines" share the physiognomical object of analysis: the body. Although practices of interpretation may differ, the physiognomical enterprise, a semiotic reading of the body, is common to the various forms of both "divination" and "rationality" in Mesopotamia.

Aristotle and the Greeks

The one text physiognomists and scholars seem to agree on as the seminal extant treatise on physiognomy is entitled *Physiognomonics* and was long attributed to Aristotle; most scholars now seem to agree that it is not in fact a text by Aristotle himself, and it is most frequently referred to as "pseudo-Aristotelian."[6] In this rather brief treatise, pseudo-Aristotle tries to establish both a methodology and a definition for a physiognomical science. He defines the science as one of particularity and individuality in the following bit of advice to future physiognomists:

> Above all it is best to base your arguments upon assertions about species and not about entire genera, for the species more nearly resembles the individual, and it is with individuals that physiognomy is concerned;

for in physiognomy we try to infer from bodily signs the character of
this or that particular person, and not the character of the whole human
race. (1240)

The logic of this advice is crucial as a fundamental basis of physiognomical thought: the relation between the particular and the general.
Pseudo-Aristotle would have us believe that physiognomy must be a
science of the particular, and as the object of its analysis, the human
body, is commonly thought of as that which is by definition most particular, this seems logical. However, he also tells us that physiognomy
should base its analyses on "assertions about species" and not "about
entire genera": in other words, the science of the particular cannot
use a large classification (genus) in order to justify its hypotheses,
but it can use a smaller one (species) to that same end. Perhaps the
mathematical odds of making a "correct" analysis are greater when
basing one's hypothesis on a more restricted classification. However,
the logic remains the same: the grounding of particular judgments
about a particular object (a human body) on observations about a
group in which the particular object can (supposedly) be classified.
The intended referents of the terms "genus" and "species" are not entirely clear here, as the present system of taxonomy and nomenclature
did not come into usage until the eighteenth century, with Linnaeus,
but one can imagine that "genus" probably refers to human beings in
general, and species to various subgroupings of people according to
race, body type, and, especially, gender.[7]

Gender distinctions are a fundamental underpinning of pseudo-
Aristotle's physiognomical system; not only does he insist on the fact
that males and females of all species are different, he also suggests
that some animal species embody some sort of essential maleness
(lions), others an equally essential femaleness (leopards). He provides
us with a list of corporeal features significant *in the male,* including
those loathesome features thought to represent femininity, but no corresponding list of features significant in the female of the species. In
short, men are both subjects and objects of Aristotelian physiognomy,
while women provide at best an obscure and decidedly unappeal-
ing point of comparison, a pejorative in the physiognomical lexicon.
The distinction between male and female, between positive and negative, is of primary importance to the budding science, according to
pseudo-Aristotle:

It is advisable, in elucidating all the signs I have mentioned, to take
into consideration both their congruity with various characters and the

distinction of the sexes; for *this is the most complete distinction,* and, as was shown, the male is more upright and courageous and, in short, altogether better than the female. (1249; emphasis mine)

The pseudo-Aristotelian treatise has always been considered the seminal work on physiognomy; it is important for the purposes of this study to recognize that, as such, it is also the site of origin of the constant link between physiognomical thought and sexism. As we shall see in later chapters, women are invariably problematized, fetishized, disparaged, and sometimes even destroyed in texts which subscribe to physiognomical notions.[8]

If we look at the physiognomical methodology proposed in the treatise, at the rhetoric of the system as opposed to its content, we find (not surprisingly) that its sole basis is one of classification and ordering. Pseudo-Aristotle proposes three basic methods for making judgments according to physical appearance (significantly omitting that of gender): zoological (comparisons between men and certain species of animals which they are said to resemble);[9] ethnological (features, both of body and character, reputed to be specific to a race or nation); and pathognomical (the emotions as they appear on the face, not in the fixed form of facial features but in transitory expression).[10] One need not be a master of rhetoric to recognize the contradictory and illogical premises on which this original physiognomical treatise is built. Perhaps the most fundamental question posed by this text, and one which is to remain a keystone for dismantling the logic of all subsequent physiognomical texts (including Lavater's), lies in the idea of physiognomy as a science of the particular, as I have mentioned above.

If we accept one of the *OED* definitions of the word *science* ("a branch of study which is concerned either with a connected body of demonstrated truths or with observed facts *systematically classified* and more or less colligated by being brought under general laws, and which includes trustworthy methods for the discovery of new truths within its own domain" [emphasis mine]), how can the very notion of a "science of the particular" have any meaning whatsoever? Indeed, it is precisely this obvious contradiction, the idea of systematizing individual personal phenomena (particularly when there are no "demonstrated truths" involved), which renders physiognomy devoid of demonstrable justification as a science. It is in this void that all subsequent physiognomists attempt to anchor their systems.[11]

While pseudo-Aristotle's treatise is the most influential and most widely appropriated by later physiognomists, it is by no means the only one of its period. However, information about specific texts on the subject and their authors is limited. Statements of general physiognomical belief can be found in, among others, both Plato (who equated physical and moral beauty, as would Lavater) and Socrates (who equated "vice and ugliness," as would Lavater).[12] What can be said is that it was the Greeks who began to attempt to elevate physiognomy to the status of a science, with generalized principles and rules.[13]

In Greece, as in Mesopotamia, the history of physiognomy is inextricably linked to that of medicine, and Pythagoras, Hippocrates, and the doctors Loxos (fourth century B.C.) and Polemon (second century A.D.) were certainly among the earliest proponents of the would-be science. Galen, himself perhaps the most widely recognized medical physiognomist, indeed asserts that it was Hippocrates who first attempted to systematize physiognomy.[14] As in Mesopotamia, medicine and physiognomy in ancient Greece shared the same general principles, including the celebrated theory of humors and their effect on both physical appearance and behavior.[15]

It is important to note in passing, however, that in spite of widespread acceptance of physiognomical doctrines such as those put forth by pseudo-Aristotle and the early medical tradition, the "science" also already had its detractors: Pliny the Elder, for example, disagreed with several of the basic tenets of Aristotelian physiognomy.[16]

Rome

In Rome, interest and belief in physiognomy continued. Cicero, for example, made at least two explicit statements of belief in the notion that a person's character was imprinted on his face by nature.[17] The most complete extant Roman treatise on physiognomy is anonymous and is thought to date from the second half of the fourth century (A.D.). As will be the case for physiognomical treatises through Lavater, the text is largely taken from earlier (mostly Greek) sources (Loxos, pseudo-Aristotle, Polemon) and would therefore seem to be as much a compilation as a truly original work.

In his 1981 bilingual (French-Latin) edition of the treatise, Jacques André explains that the anonymous author employs three basic physiognomical methods, derived in large part from pseudo-Aristotle and familiar to the student of physiognomy of any period: the anatomical (a list of the various body parts and their meanings, such as we shall see in most later physiognomists), the zoological, and the ethnological. As André points out in reference to the ethnological principle (and this holds true for the zoological as well), the rhetoric of such analysis is that of syllogism: if a man looks like a hog, and if hogs are known to be gluttonous and lazy, then it follows that the man is also gluttonous and lazy.[18] The physiognomical syllogism, often attributed to the Neapolitan Giambattista della Porta of the late sixteenth century, was in fact a fundamental rhetorical underpinning of physiognomical thought as early as the anonymous Roman treatise, who indeed had inherited it from pseudo-Aristotle.

Of greatest interest to me in the anonymous treatise, as a prefiguration of many of the eighteenth- and nineteenth-century literary manifestations of physiognomical thought we shall see later, is the emphasis placed on physiognomical difference between the sexes. Roman physiognomy, echoing pseudo-Aristotle, is (among other things) a discourse of sexism, as indeed theories of biological determinism almost invariably are. André, appropriately, remarks that physiognomy in the ancient world (and I would argue that this remains true throughout the history of physiognomy) is addressed solely to men.[19] Given the historical context in which such writings were produced, it would be perhaps anachronistic (and indeed redundant) to characterize ancient physiognomy as "sexist" solely on the basis of its male-only audience, or even perhaps on the basis of its primitive definitions of sexual difference. More significantly, the very first section of the treatise defines the masculine and feminine characters as traits which are innate but which do not necessarily correspond to anatomical sex: the feminine can be found in men, and vice versa.[20] This is a curious and important concept which does not fit into our contemporary scheme of distinguishing the anatomical (sex) from the cultural (gender). Early physiognomy seems to describe and perhaps to have invented an essentialist category which lies somewhere between the two.[21] Not surprisingly, the character type described as female is considerably less appealing than the male: whereas the male is, among other things, "impulsive," "generous," "upstanding," and "magnanimous," the female is "envious," "pitiless," "fearful," and "hypocritical." The reader is tempted to wonder why the author of the treatise didn't simply state that positive qualities can be described as masculine, and negative as feminine. As André remarks, supposedly "feminine" characteristics are pointed out as flaws in a man (as signs of effeminacy and, significantly, of debauchery), but never, although it is certainly given as theoretically possible, are positive, "masculine" traits ascribed to a woman.[22]

The Middle Ages and the Renaissance

Translations of both the ancient treatises on physiognomy and the important Arabic works on the subject began in the twelfth century. These translations renewed interest in the would-be science, and soon sparked contemporary writings on the subject, mostly derivative of the Greek and/or Arabic traditions. Medieval physiognomy was characterized by the same seemingly contradictory approaches present in its predecessors: a more rational, "scientific" method and a divinatory, supernatural method, often based on astrology. It also continued to be linked with the theory and practice of medicine, and consequently with the notion of the humors and temperament, as well as with more forward-looking elements such as a systematic study of anatomy.[23] Interestingly, what we see today as a contradiction, the relation (or lack thereof) between theories of astrology, temperaments, and medicine, was often seen as merely the basis for a single system, of which physiognomy was the first, and most concrete, step. Roughly speaking, the logic was as follows: the stars influence the disposition of physiognomical temperament in a given individual, and the temperament expresses itself through the outer body, which is in turn "read" physiognomically.

The Renaissance, with its renewal of interest in Greek and Roman texts, brought attention once again to physiognomy as such. Only at the very end of the sixteenth century, however, with the dawn of "modern" science, does one begin to see the genesis of a "natural" physiognomy which distances itself from affiliation with astrology and divination.[24] The best-known physiognomist of the period is Giambattista della Porta (1535?–1615), author of *De humana physiognomia* (1586), a widely translated and circulated treatise whose content amounts for the most part to a new presentation and reinterpretation of both the classical and medieval texts on physiognomy and medicine.

While he continues to associate physiognomical theory and practice with the quasi-medical notions of temperament and the humors, Porta's work is better characterized by his adherence to the Aristotelian tradition of zoological physiognomy, and includes many remarkable and eerie engravings comparing human heads with those of particular species of animals.[25] Porta's work is characterized as well by a clearly more skeptical view of the astrological method of physiognomy which dominates the work of so many of his predecessors. As for the other various divinatory traditions, Porta takes his stand with the heading of the first book of his work: "Che molte scienze divinatrici siano vane, false, e pernitiose, e quanto sia grande l'eccellenza della Fisonomia, come nata da principi naturali."[26] As Courtine and Haroche have said, Porta's work is at bottom a triumph of Aristotelian, "natural" physiognomical principles over those of astrology[27]; it indeed represents a significant and definitive transition from divination to the more rational, empirical methods introduced in the seventeenth century.[28]

The Seventeenth Century

While seventeenth century physiognomy per se never really caught up with contemporaneous advances made in the natural sciences and medicine, it was nonetheless clearly marked by a movement toward a more scientific discourse, and a new conception of the signifying powers of the body.

Perhaps the most influential figure in the history of seventeenth-century physiognomical thought is Descartes, who is of course rarely identified as a physiognomist at all. Descartes's contributions to the field were both direct and indirect. His *Discours de la méthode* (1637) defined the burgeoning principles of rationality, order, and method, clearly the antitheses of the divination and emotionalism which had long been associated with physiognomy. It is important to note that it was in an intellectual climate dominated by the Cartesian *episteme* that physiognomical thought experienced yet another sort of renaissance in the mid-seventeenth century.

Descartes's more direct influence on physiognomy came from his last work, *Les Passions de l'âme* (1649). In this treatise, Descartes gives both physiological and philosophical definitions of the relation between soul and body. The Cartesian "passions," as both physiological and affective phenomena, are the *point de rencontre* of the otherwise disjointed body and soul, and are therefore the most important object of study. In identifying the pineal gland as the locus of the passions, Descartes clearly demonstrates his desire to inscribe his treatise into a tradition of anatomical, physiological inquiry. Indeed, as he says in a prefatory letter, "mon dessein n'a pas esté d'expliquer les Passions en Orateur, ny mesme en Philosophe moral, mais seulement en Physicien."[29]

Largely influenced by contemporary theories of circulation (that of Harvey being the most important) and thus, indirectly, by the ancient theories of the humors, Descartes's physiological theories are of less interest to a student of physiognomy than are his notions of the relations between body and mind and body and soul. The Cartesian passions are legible through the body, and it is this notion which anchors Descartes's theories to explicitly physiognomical thought. No longer subscribing to a theory by which the body is merely a text comprising corporeal signs written by some supernatural power, Descartes notes the importance of movement, and of the voluntary nature of facial expression:

> Il n'y a aucune Passion que quelque particuliere action des yeux ne declare . . . des actions du visage, qui accompagnent aussi les passions . . . elles ne semblent pas tant estre naturelles que volontaires. Et generalement toutes les actions, tant du visage que des yeux, peuvent estre changées par l'âme, lors que voulant cacher la passion, elle en imagine fortement une contraire: en sorte qu'on s'en peut aussi bien servir à dissimuler ses passions, qu'à les declarer. (147, article 113)

What is particularly interesting in the passage above is Descartes's explicit admission of the possibility of physiognomical fraud on the part of the object of study: the human body is indeed not an immobile, involuntary object, and man is quite capable of altering his face in order to deceive. As we shall see in studying later theoreticians of related questions, it is essential that any rational form of physiognomical thought keep this fact as one of its most fundamental tenets.

The notion of the passions, and of their expression, will be central to Le Brun's work on methods of painting human expression later in the seventeenth century.[30] Le Brun, director of the Academy of Painting and Sculpture under Louis XIV, gave a famous discourse, known as "Conférence sur l'expression des passions," on the subject of the passions, their expression, and the method of drawing these expressions before the academy (1668); various versions and abridgements of this "conference," profusely illustrated by strangely beautiful, if sometimes grotesque, exemplary engravings, were translated and circulated throughout the late seventeenth and early eighteenth centuries.[31]

In a sense, Le Brun's work can be viewed as both a restatement, and a sort of elaborate *mise-en-pratique,* of Descartes's ideas, including the most basic physiognomical belief:

> l'expression est . . . une partie qui marque les mouvements de l'âme, ce qui rend visibles les effets de la passion.
> . . . tout ce qui cause à l'âme de la passion, fait faire au corps quelque action.

> Comme il est donc vrai que la plus grande partie des passions de l'âme produisent des actions corporelles, il est nécessaire que nous sachions quelles sont les actions du corps qui expriment les passions . . .[32]

Indeed, even Descartes's notion of the pineal gland as the locus of passions is to be found in Le Brun, albeit somewhat modified:

> Quoique l'âme soit jointe à toutes les parties du corps, il y a néanmoins diverses opinions touchant le lieu où elle exerce plus particulièrement ses fonctions. Les uns disent que c'est une petite glande qui est au milieu du cerveau . . . D'autres disent que c'est au coeur, parce que c'est en cette partie que l'on ressent les passions; et pour moi, c'est mon opinion que l'âme reçoit les impressions des passions dans le cerveau, et qu'elle en ressent les effets au coeur. (96)

However extensive and interesting Le Brun's debt to Descartes may be,[33] the real interest for a student of physiognomy lies in Le Brun's expanding upon Descartes's notion of the legibility of the passions. Descartes is most concerned with describing the passions themselves ("La Tristesse est une langeur desagreable, en laquelle consiste l'incommodité que l'ame reçoit du mal, ou du defaut que les impressions du cerveau luy representent . . .") and the physiological "causes" and symptoms of these passions ("En la Tristesse, que le poulx est foible & lent, & qu'on sent comme des liens autour du coeur, qui le serrent, & des glaçons qui le gelent, & communiquent leur froideur au reste du corps . . ."). Le Brun reproduces, word for word and without acknowledgment, some of Descartes's definitions of the passions (including those quoted above). He is, however, much more concerned with defining the visible facial effects of these physiological phenomena.[34] As a painter, it is his goal to teach young painters how to represent the facial expressions of admiration (this rather unexpected choice as a primary "passion" is an obvious legacy from Descartes), joy, sadness, etc. Obviously, in order to draw these expressions, one must know how to recognize them. Le Brun's work is important for many reasons, not the least of which is the practical, "how-to" aspect which Lavater aspires to (and fails to achieve) a century later.

Most of the editions of Le Brun's work are more notable for their copious illustrations of heads expressing the various passions than for their contribution to the philosophy or "science" of physiognomy. Both Le Brun's practice and his theory strongly suggest that only through visual representation can certain expressions be rendered:

[we must] observe the material sign, which changes all the others, and increases or lessens their force and power; which cannot be well understood or demonstrated but by a Figure.[35]

In making this statement, Le Brun is of course implying that verbal representation is inadequate as a medium for rendering the expression of the passions. If, as Descartes showed, the passions as psychophysiological phenomena can be described through language, as esthetic or visual phenomena they cannot. This skepticism about the power of verbal language to represent the body is indeed as old as physiognomical thought itself, having been expressed as early as the Mesopotamian treatises on the topic.[36] What Le Brun's work does, appropriately enough, is illustrate the need for illustration. Although other works of physiognomy (notably that of Porta, who may or may not have inspired Le Brun) had been illustrated, editions of Le Brun's work were often presented as a series of plates with accompanying text, rather than vice versa.[37] As I shall show with Lavater, faith in the visual over the verbal is a fundamental tenet of physiognomical thought, particularly in the acutely language-conscious eighteenth century. My discussion of literary works will suggest (albeit obliquely) that the apparent inability of verbal language to represent and explain the human body is at the heart of the tension created by physiognomical themes in the novel.

Le Brun's most notable contribution to physiognomy per se is his emphasis on the importance of facial expression[38] and, specifically, his belief that, of all the body parts, the eyebrow (as opposed to the eye itself) is the most expressive:

comme nous avons dit que la glande qui est au milieu du cerveau est le lieu où l'âme reçoit les images des passions, le sourcil est la partie de tout le visage où les passions se font mieux connaître, quoique plusieurs aient pensé que ce soit dans les yeux. (99)

This seemingly minor contribution to physiognomical method implies nothing short of a revolution in physiognomical thought which takes place in the seventeenth century with Descartes and his successors. Le Brun's privileging of the eyebrow is predicated on the eyebrow's ability to express emotion through movement. Movement is indeed what expression is based on: after Descartes's "secularization" of the body as an organic, mechanistic whole, it can no longer be read as an immutable text inscribed by divine or occult powers.[39] The body must be read as self-referential: that is to say, the body (through facial expression, for Le Brun) reflects the inner state of the individual at a particular moment, and not some eternal essence of his being. Furthermore, individual features in Le Brun are read not as signifiers in and of themselves, but rather as parts of a whole, in relation to other features and to their own appearance in repose or as part of another expression.[40] It is important to note that Le Brun's famous "heads" demonstrating the various passions are devoid of any distinctive features (and even of hair): they are intended to be as "pure" expressions of the passions as possible, free of any semiotic significance other than that of the passion in question.

This approach to the question of the legible body, born of the growing scientific rationality and awareness of the functions of the body as a machine which characterizes the period (as exemplified by Descartes), is a radical departure from all earlier physiognomical theory, which tended to privilege the stable, immobile parts of the body as more reliable indices. As Courtine and Haroche have astutely remarked about this shift, it was as if a model by which the body was perceived as volatile spoken language ("parole") were substituted for the previous model of analogy with an unchanging written text.[41]

Equally significant is the shift in Cartesian-inspired physiognomy from the traditional rhetorical model of analogy to that of causality: no longer does the body signify because of a resemblance, direct or indirect, to anything else. With Descartes and Le Brun, the physiognomical syllogism is no longer in effect. Instead, the body signifies because of its own logical, organic, and physiological principles. The expression of the passions is, according to the Cartesian understanding of the body, nothing more than the effect of (more or less) easily definable physiological conditions.[42] This important change in both method and rhetoric greatly "modernizes" physiognomical thought and reaffirms, yet again, its link to ongoing developments in medicine. Interestingly, the next important moment in the history of physiognomy will not come until a century later, with Lavater, and represents more of a regression than a progression vis-à-vis Descartes and Le Brun.

Although Descartes and Le Brun are perhaps the most suggestive and important figures in physiognomical thought in a general sense in the seventeenth century, neither is a theorist of physiognomy as such. Before moving on to a discussion of physiognomical themes in Marivaux, however, it would be useful to discuss briefly the most widely known physiognomist of the period. Marin Cureau de La Chambre, physician to Louis XIV, was the author of a treatise on the passions heavily inspired by Descartes's *Les Passions de l'âme*—*Les Charactères des passions* (1653)—which in turn may have influenced Le Brun. Cureau insists on his goal of combining medicine, morality, and politics in his work ("je prétends mettre ce que la Medecine, la Morale & la Politique ont de plus rare & de plus excellent"). A somewhat later work is more explicitly physiognomical, in its title if not in its content: *L'Art de Connoistre les Hommes, où sont contenus Les Discours Préliminaires qui servent d'Introduction à cette Science* (1667). The title sounds familiar to anyone who knows Lavater, the title of whose own work was almost certainly influenced by it.

Cureau's physiognomical beliefs are fairly typical of the period in which they were written: the body shows "inclinations" which are in turn the product of "temperament" (as defined by physiology since the Greeks). These beliefs are accompanied by an expression of grave doubt concerning, if not an actual refutation of, physiognomical divination in the forms of chiromancy and metoposcopy. Not surprisingly, Cureau's logic is, in principle at least, predicated on the Cartesian model of cause and effect ("parce qu'il n'y a point de rapport de cette nature que celuy de la cause à son effet, ou de l'effet à la cause . . .").

What is more interesting to me than these fairly predictable notions is Cureau's privileging of *air*, a concept which will become important in my subsequent discussions of physiognomy as a literary theme. According to Cureau, the temperament gives an *air*, a sort of general appearance or impression. In his "Avis nécessaire au lecteur," which serves as a preface to *Les Charactères des passions*, Cureau defines the very first rule of physiognomy as follows:

> La première [règle] est fondée sur les Charactères des Passions, des Vertus, & des Vices; et fait voir que ceux qui ont naturellement le mesme Air qui accompagne les Passions ou les Actions à des Vertus & des Vices, sont aussi naturellement enclins aux mesmes Passions & aux mesmes Actions.[43]

Aside from his reliance on a logic akin to that of physiognomical syllogism,[44] the most striking thing about Cureau's "rule" is its faith in the reliability of *air*. In a noteworthy break with physiognomical tradition, Cureau refers not to specific corporeal signifiers, but rather to a general appearance and the subjective impressions it conveys. Although he fails to define what *air* entails, we can safely surmise that it is a vague amalgam of beauty (or at least "regularity" of features), posture, gesture, and expression. *Air*, as opposed to morphology, almost certainly allows for movement and expression as objects of interpretation, and is thus very much in keeping with the seventeenth-century physiognomical view we saw above. The interpretation of *air* is furthermore by necessity much freer and more impressionistic than that demanded by fixed corporeal features. The implications of a physiognomy which emphasizes *air* will become obvious in our discussion of Marivaux's *La Vie de Marianne*, an early-to-mid-eighteenth-century text (1731–41) whose treatment of physiognomy is decidedly reminiscent of the last half of the previous century.

A second notable aspect of Cureau's physiognomy, for my purposes, is the emphasis he places on the difference(s) between the sexes as a fundamental principle of physiognomy. In *Les Charactères*, Cureau posits sexual difference as the third of his five "règles générales":

> La troisième est fondée sur la beauté des sexes, & montre que les hommes qui ont quelque chose de la Beauté féminine, sont naturellement effeminez, & que les femmes qui ont quelque chose de la Beauté virile, participent aux inclinations des Hommes. ("Avis")

While such ideas are certainly not novel in either their content or their rhetoric, and indeed, as we have seen, can be traced back to the seminal classical treatises on physiognomy, it is important to note that they persisted almost without modification through the purportedly increasingly scientific seventeenth century and well beyond. In keeping with tradition, Cureau believes not only in an absolute definition of gender as based on sexual (anatomical) difference, but also in the enigmatic and inferior nature of women:

> Il faut maintenant examiner . . . la Femme. Mais que cette entreprise est difficile! qu'elle est perilleuse! . . . sans doute, s'il y a quelque certitude dans le raisonnement, si les principes que la Nature a versés dans notre Ame pour la connaissance de la vérité ont quelque chose de solide, il faut de necessité qu'il n'y ait pas une de toutes les parties qui sont nécessaires pour former la Beauté de la Femme, qui ne soit la marque d'une inclination à quelque vice.[45]

Even if Cureau makes the gesture of specifying that in actual practice, women are more likely to be virtuous than men, the notion that "vice" is a component inherent in the essential nature of woman is not contradicted. In an echo of earlier physiognomical treatises (one is reminded of the anonymous Roman work, for example), negative qualities alone are ascribed to women ("foible, timide, soupçonneuse, rusée, dissimulée, menteuse, aisée à offenser, mobile, légère, pitoyable, babillarde . . .").[46]

There are several elements in the exemplars of seventeenth-century physiognomical thought which will be germane to my analysis of texts by Marivaux, particularly the novel *La Vie de Marianne*. I hope to have sketched at least a brief outline here of the misogyny linked to much of the physiognomical tradition up until, and including, the late seventeenth century. Physiognomical thought very often entails an essentialist (and hierarchical) definition of sexual difference, in other words the belief that anatomical difference necessarily implies moral, intellectual, and behavioral difference as well. It is obvious how and why such a notion might grow out of, and remain intertwined with, notions of the body as signifier. This seemingly inevitable relation between physiognomy and its corollary, sexism, will be a constant, if not always explicit, underpinning of most of my readings of literary texts in the chapters to follow. Nowhere will this be more evident than in Marivaux, who may or may not qualify for the term "misogynist" (there has been considerable discussion about this), but in whose texts questions of sexual difference and of the "eternal feminine" are at the forefront.

To speak somewhat more historically, we shall see that *La Vie de Marianne* illustrates and, perhaps, questions some of the currents in late-seventeenth-century physiognomical thought, the most recent important moment in the evolution of physiognomical thought when *Marianne* was written (1731–41). Nowhere in Marivaux will we find the morphological analysis so essential to the post-Lavater Balzac; instead there is an emphasis, recalling that of Cureau, on *air* and on physiognomical syllogism (if aristocrats are gracious and refined and Marianne is gracious and refined, then Marianne is an aristocrat). Furthermore, Le Brun's preoccupation with expression and the mobile features, which Lavater will later define as "pathognomy," suggested the importance of physiognomy in the arena of social politics. For the first time in the history of physiognomy, emphasis is placed on features which might well be controlled by the individual will, thus allowing for the possibility of deception. This revelation is mirrored in French literature of the seventeenth and eighteenth centuries by the theme of the mask, and all of the tortured plots it produces, particularly in the subgenre Peter Brooks has called "the novel of worldliness" (to which *Marianne* belongs).[47] The importance of the theme of the mask in Marivaux has been amply demonstrated; in the following chapter, I hope to show, among other things, that the mask is indeed one among several physiognomical themes of central importance in Marivaux's work.

Chapter Two

Marivaux, *le masque,* and *le miroir*

En vous peignant ces hommes que j'ai trouvés, je vais vous donner le portrait des hommes faux avec qui vous vivez, je vais vous lever le masque qu'ils portent. Vous savez ce qu'ils paraissent, et non pas ce qu'ils sont. Vous ne connaissez point leur âme, vous allez la voir au visage, et ce visage vaut bien la peine d'être vu; ne fût-ce que pour n'être point la dupe de celui qu'on lui substitue, et que vous prenez pour le véritable . . .

—Marivaux

La Vie de Marianne ou les aventures de la comtesse de *** is a memoir-novel written in eleven installments from 1731 to 1741 by Pierre Carlet de Chamblain de Marivaux. As a very young child, the fictional memorialist was the sole survivor of a carriage accident. Because she was found pinned beneath the corpses of both a noblewoman and her domestic servant, her class identity is not verifiable. In fact, as the names of the "gens de condition" in the carriage are unknown (they are thought to be foreigners), even though the child is presumed to be theirs, she remains a nameless and classless enigma. Cared for and named by a local priest and his sister, Marianne grows up to be a young woman of unusual beauty, virtue, and refinement. Circumstances lead Marianne to Paris, where she soon finds herself on her own. Her subsequent adventures and misadventures in the merchant-class, conventual, and aristocratic milieux all revolve around the central question of her pitiable lack of familial, and therefore class, identity. Although this clearly puts Marianne in a position of great disadvantage in society, she manages to procure benefactors and well-wishers through her remarkable personal characteristics. It is not, however, merely the fact of her beauty and grace which wins the hearts of her admirers; it is rather the innate aristocracy that these qualities seem to demonstrate. Marianne's person itself becomes the "proof" of her noble blood, in the absence of all sources of verification.

The primary interest of this particular novel for a student of physiognomy is simple: Marianne is a character in whom essence seems

33

to be directly accessible through appearance. It is the extent to which this phenomenon is at issue in the novel that renders it particularly relevant to my discussion. *La Vie de Marianne* is a novel in which both the *histoire* and the *récit* are motivated by the physical appearance of the protagonist. Physiognomy is both explicitly and implicitly at issue throughout the book, and Marianne the narrator proves to be as interesting a physiognomist as Marianne the character is an object of physiognomy. She provides her readers with many pathognomical and physiognomical insights about other characters, as well as several excellent examples of lengthy *portraits,* the eighteenth-century set pieces which evolved from the seventeenth-century Theophrastean *caractère.*

Although the raw material for an analysis of physiognomy in this novel is more than sufficient within the work itself, it becomes even more complex and intriguing when read in conjunction with Marivaux's periodical writings, themselves veritable manifestos of the "science" of reading true character from often deceptive appearances. Because the periodicals are more explicitly theoretical than *La Vie de Marianne* about the important questions pertinent to a discussion of physiognomy, it would perhaps be helpful to use them as the starting point for a discussion of physiognomy and related concerns in Marivaux.

The Periodicals

Marivaux's first periodical was *Le Spectateur français,* patterned after Addison and Steele's famous *Spectator* in England. The journal was published in twenty-five separate installments (*feuilles*) between 1721 and 1724, and contains a variety of literary genres from anecdote to *conte* to essay to literary criticism. In 1727, Marivaux published the seven *feuilles* of a second journal, *L'Indigent philosophe,* which is of a more provocative, more cynical, less moralistic tone than the *Spectateur. Le Cabinet du philosophe,* the third and last of Marivaux's journals, was published in 1734, in eleven *feuilles.* Like *Le Spectateur français,* it comprises pieces belonging to a variety of genres including fairy tales, theater, and essays.

While it is true that Marivaux is better known for his theater and two novels (*La Vie de Marianne* and *Le Paysan parvenu*) than for these periodicals, it is also true that they contain some of the clearest expressions of the thought behind themes that recur with obsessive regularity in his works of fiction. Nowhere is Marivaux the thinker more suggestive than when assuming, in the different periodicals, the guise of the observer of and commentator on his fellow man.

The periodicals, for all the differences which separate them, have a common theme and object of study: mankind. Following in the tradition of the *précieux* and the moralists as well as that of Addison and Steele, the periodicals seek to inform, as well as to entertain, their readers concerning the character and behavior of their fellow humans. Marivaux's study of man in his periodicals can aptly be called a physiognomical one in that the truth of man's moral, inner self is read from, and against, the superficial appearance of his social, exterior self.[1] The fundamental agenda of the semiotic system of Marivaux's observations is that to be found in almost any physiognomical treatise: to educate readers about the relationship between what they observe in their neighbors (social signifiers, including physical appearance) and the truth behind this surface (the signified, "true" character).[2] However, as is equally true of most physiognomy, Marivaux's "science" is an impressionistic, inexact one, neither based on quantifiable data nor systematized in any coherent form.

Marivaux's attempts to interpret his fellow men analyze a much broader set of signifiers than merely those of the body itself, and tend to emphasize those of social comportment and discourse. As for the instances of corporeal analysis, Marivaux's physiognomy reflects seventeenth-century influences such as Descartes and especially Le Brun in its tendency to emphasize the pathognomical (the expressions of the mobile features) rather than the truly physiognomical (the configuration of the fixed features). Marivaux uses the whole of the social persona, including acts, words, and gestures, to constitute man's social self; indeed, in an essay which postdates the periodicals (published in the *Mercure de France* in 1744), "Réflexions sur l'esprit humain," Marivaux tells his readers that it is society itself which teaches us "la science du cœur humain":

> Elle n'a pas . . . ses professeurs à part, à peine suffiraient-ils pour vous en donner la plus légère idée, et rien de ce que je dis là n'en ferait une connaissance inévitable. C'est la société, c'est toute l'humanité même qui en tient la seule école qui soit convenable, école toujours ouverte, où tout homme étudie les autres, et en est étudié à son tour; où tout homme est tour à tour écolier et maître. (476)

Unlike his successor Lavater, Marivaux purports to teach this science of reading men not so as to propagate greater love among them, but rather so as to help them protect themselves from each other. Marivaux's view of his society is at least somewhat cynical, and his physiognomy a defensive one. Comparison of the content of the three periodicals reveals a consistent, if not obsessive, worldview. While Marivaux's concept of natural, essential man is not expressed in a clear fashion in his periodical writings, his ideas about social man, about man-in-society, are unambiguous.[3] According to Marivaux, social man is obliged, by virtue of the fact that he engages in close interaction with others, to adopt various personae which have greater or lesser degrees of relation to his "real" character. Obviously, the personae which are furthest from this essential character and therefore most artificial are of the greatest interest to both the social scientist and the moralist for their value as objects of analysis. In keeping with common eighteenth-century usage, Marivaux uses the metaphor of the mask to define this phenomenon. All men in society are mask wearers, showing the "face" which will be the most pleasing to the spectator of the moment, and thereby hiding the true "face."[4]

In his groundbreaking study *Marivaux's Novels: Theme and Function in Eighteenth-Century Narrative*, Ronald Rosbottom defines mask wearing as compromise, as a sort of necessary evil in the society described by Marivaux:

> Neither honesty nor hypocrisy can ensure the successful functioning of society as Marivaux defines it. A compromise must be made that will permit people to be sincere, without hurting others, and sociable. Since everyone is possessed of a well-defined ego, and since he wishes to present himself successfully in society, compromise is inescapable. Some form of social mask becomes incumbent upon all participants in a given society. Each individual, in an attempt to safeguard his ego, must evolve a science of defense. Marivaux realized the unfortunate consequences of such a world view: it would mean the denial of his belief in the necessity of sincerity in human relations. But, given the social context of the period, it was the only door left open to the Modern moralist.[5]

Rosbottom's analysis of Marivaux's recognition of the inevitability of moral compromise in a successful society is astute; however, I would emphasize somewhat more than he Marivaux's awareness of the unhappy consequences of such an inevitability.

In the oft-cited anecdote which concludes the first *feuille* of the *Spectateur*, the narrator recounts that in his youth he fell very much in love with a beautiful young woman precisely because she was "simple" and had no awareness of her own charms ("Quel plaisir! disais-je en moi-même, si je puis me faire aimer d'une fille qui ne souhaite pas d'avoir des amants, puisqu'elle est belle sans y prendre garde . . ."). In short, she was a woman who wore no mask (it is not without significance that the seeming absence of a mask is signaled by a seeming lack of concern for the physical person, the literal "face"). The young lover's illusions are shattered when he discovers the young woman at her mirror, practicing and perfecting the very same gestures and expressions he had thought to be guarantees of her candor ("il se trouvait que ses airs de physionomie que j'avais cru si naifs n'étaient, à les bien nommer, que des tours de gibecière . . ."). It is of primary importance to an understanding of Marivaux's social science to note that this very first anecdote of all his periodicals is a dramatic mise-en-scène of the physiognomical *tromperies* that take place in society; it is also important to note that the perpetrator of this deceit is a woman, and a woman before a mirror—themes whose significance we will discuss below.[6]

In a remark about this anecdote, Rosbottom suggests that Marivaux resigns himself to a relativistic stance vis-à-vis the trickeries of social man:

> Marivaux set the tone of the *Spectator* in the first number with the anecdote about the girl who was practicing her charms before a mirror. He had decided to accept a stance of moral relativism after this event: he will no longer judge men, but rather analyze them and their actions in social terms . . .[7]

The anecdote does indeed set the tone for the periodical, in fact for all of Marivaux's study of man, but I would argue that this tone is not entirely one of relativism and resignation. Marivaux maintains a moral stance, albeit often implicitly, toward social deceit throughout his periodicals and even in his novels. The narrator of the anecdote himself makes it clear that this experience started him on his career, not as a neutral and objective observer, but as a cynical, judgmental, and indeed misanthrophic, one, concluding the *feuille* with the following revelatory statement: "Je sortis là-dessus, et c'est de cette aventure que naquît en moi cette *misanthropie* qui ne m'a point quitté, et qui m'a fait passer ma vie à examiner les hommes, et à m'amuser de mes réflexions."[8]

It is clear that the narrator is describing a primal scene of loss of social innocence, and that with this loss of innocence comes a certain measure of cynicism. He says that the spectacle of social relations and their hypocrisies will "divert" and "amuse" him, but implies that his laughter will be one of scorn and disapproval. His study of man will serve to justify his misanthropy by ever lengthening his catalogue of social hypocrisies.

If society is composed of mask wearers, and if it is the role of the social scientist or "spectator" to unmask them, what exactly can comprise the masks? Marivaux puts forth the rather revolutionary idea (totally contradicted by the ethos presented in *La Vie de Marianne*, as we shall see later) that a spectator must read beyond *all* appearances, including those which society and culture ordain, such as social class:

> Dans un domestique, je vois un homme; dans son maître, je ne vois que cela non plus, chacun a son métier; l'un sert à table, l'autre au barreau, l'autre ailleurs: tous les hommes servent, et peut-être que celui qu'on appelle valet est le moins valet de la bande; c'est là tout ce que le bon sens peut voir là-dedans, le reste n'est pas de sa connaissance, et dans l'état où je suis, on n'a que du bon sens, on perd de vue les arrangements de la vanité humaine. (*L'Indigent philosophe*, Sixième Feuille, 303)

While it would be ludicrous to ascribe what we would consider to be truly democratic views to Marivaux, the above passage does demonstrate the breadth of his belief in the necessity of reading beyond all that is superficial, including codified social status.[9] The ways in which men group themselves into hierarchical systems would seem to be of no interest to the commentator of essential human character (he who possesses what is characterized above as "le bon sens"), who sees them as merely "arrangements de la vanité humaine." In fact, however, this apparent dismissal of social systems belies the content of this and the other periodicals somewhat. While the spectator is not duped by exterior means of bolstering human vanity (including rigid systems of social class), he does spend his time uncovering various instances of these "arrangements de la vanité." Indeed, if the periodicals can be said to have a common theme, it would be that of the hypocrisies which result specifically from excessive vanity. While the masks might appear to be different, they are all made of the same substance. Just as in the "primal" anecdote of the young woman at her mirror in the *Spectateur*, all of the instances of unmasking in the periodicals involve distinguishing those personal characteristics learned before the mirror of petty personal vanity.

"Le Voyageur dans le Nouveau Monde"

It would perhaps be of interest at this point in my study to look at a specific example of some of the generalities I have been discussing. "Le Voyageur dans le Nouveau Monde" is a sort of fairy tale told in the sixth through eleventh *feuilles* of *Le Cabinet du philosophe*. It is by far the clearest and most inventive of Marivaux's attempts to illustrate his theories about the hypocrisies of social commerce.

"Le Voyageur," as its title would indicate, is a travelogue—with a twist. In it, a man recounts an incident from his youth which led him to visit a strange land. The young man, after having been betrayed by both his best friend and his mistress, flees France. In the course of his travels, he meets an older man, also French, who befriends him and offers to take him on a voyage to a land where men are genuine and "ne se contrefont point." After some rather mysterious remarks on the part of the older man and some required reading for the younger (*L'Histoire du cœur humain*), the pair set out on a sea voyage. After some time at sea, during which the younger man has no sense of the geography of their excursion, they disembark and find themselves in a country which looks very much like France and whose inhabitants speak French. When the young man asks his guide if they are in fact in France, the guide responds: "Non pas dans la France que vous connaissez . . . mais dans celle de ce nouveau monde où je vous mène, et qui est exactement le double du nôtre" (397). The guide further warns him that although the inhabitants of this country are, as promised, "vrais," they will *try* to be false and therefore the young man must use his knowledge of the insincerity of human interaction to see the truth ("méfiez-vous d'eux comme s'ils étaient faux; servez-vous avec eux des lumières que vous avez acquises . . . vous vous apercevrez bien un peu des efforts inutiles qu'ils font d'abord pour se déguiser" [398]).

Upon arrival in the port city, the young man immediately encounters a man who exactly resembles an acquaintance and who seems to recognize him. The difference in this double is the fact that he is boastful, ambitious, and smug—qualities the young man never noticed in his friend before. After meeting the double's fiancée, who expresses interest in his person in no uncertain terms, and having observed several remarkably rude verbal exchanges at a social gathering, the bewildered young man and his guide leave the town and travel to what appears to be Paris (or, as the guide would have it, the exact double of Paris). After a series of encounters with people who exactly resemble those whom he knew quite well, but whose behavior and discourse is completely uncharacteristic, the young man finally demands an explanation for the bizarre mirror world he has entered. His guide informs him that he is in fact in the "real" France, but that it is indeed also a strange land to him because he has acquired the secret of seeing things as they really are, of not being duped by appearances:

> Vous n'aviez jamais vu d'hommes vrais: je vous avais promis de vous en faire voir, et vous les avez vus. Ce ne sont pas d'autres gens que ceux de notre monde, j'en conviens; mais ils ne sont pas moins nouveaux pour vous, puisque vous les avez pris pour des hommes d'une espèce différente, et que vous n'en avez reconnu que la physionomie, et non pas le caractère. Les voilà tels qu'ils sont, au reste; et à présent que la lecture des livres que je vous ai donnés, et que les réflexions que vous avez faites en conséquence, vous ont appris à connaître ces hommes, et à percer au travers du masque dont ils se couvrent, vous les verrez toujours de même, et vous serez le reste de votre vie dans ce Monde vrai, dont je vous parlais comme d'un monde étranger au nôtre . . . (419)

"Le Voyageur dans le Nouveau Monde" is an ingenious *roman d'initiation* in which the young man is taught the necessity of accurate character analysis in the world (it is important to remember that the tale begins with his failure to assess the characters of his best friend and mistress) by a seemingly fantastic illustration of the vast gap between appearance and character. Indeed, the tale is the story of an initiation into physiognomy. The following is a compelling statement of the fact that the unmasking described is in fact the physiognomical act:

> De sorte qu'en vous peignant ces hommes que j'ai trouvés, je vais vous donner le portrait des hommes faux avec qui vous vivez, je vais vous lever le masque qu'ils portent. Vous savez ce qu'ils paraissent, et non pas ce qu'ils sont. Vous ne connaissez point leur âme, *vous allez la voir au visage, et ce visage vaut bien la peine d'être vu;* ne fût-ce que pour n'être point la dupe de celui qu'on lui substitue, et que vous prenez pour le véritable . . . (389; emphasis mine)

In trying to explain to his readers how he came to read people so accurately, the narrator tells us:

> Leur naïveté n'est pas dans leurs mots (j'ai peut-être oublié d'en avertir): elle est dans la tournure de leurs discours, dans l'air qu'ils ont en parlant, dans leur ton, dans leur geste, même dans leurs regards: et c'est dans tout ce que je dis là que leurs pensées se trouvent bien nettement, bien ingénument exprimées; des paroles prononcées ne seraient

pas plus claires. Tout cela forme une langue à part qu'il faut entendre, que j'entendais alors dans les autres pour la première fois de ma vie, que j'avais moi-même parlé quelquefois, sans y prendre garde, et sans avoir eu besoin de l'apprendre, parce qu'elle est naturelle et comme forcée dans toutes les âmes. Langue, qui n'admet point d'équivoque; l'âme qui la parle ne prend jamais un mot l'un pour l'autre . . . (401)[10]

This tale would almost seem to belong to the fantasy genre, and indeed the young man wonders at one point if his guide is not some kind of magician. However, when the young man realizes that he has been seeing the true character and hearing the thoughts of those he encounters not through any sorcery, but merely through his own heightened perception, the distinction between fantasy and reality breaks down. The young man has not taken a fantastic voyage, and he has not been endowed with powers that, strictly speaking, can be called supernatural. He has been endowed with the powers of the observer, the spectator, the physiognomist. Physiognomy the "science" is defined by the guide in an earlier passage when he says that the young man has previously recognized only "la physionomie, et non pas le caractère" and thereby makes the all-important point that "physiognomonie" (although he does not use the word, which had yet to be popularized by Lavater) means seeing beyond "physionomie." These are powers which are within the reach of most intelligent people but which blur the distinctions between fantasy and reality, and constitute a sort of extrasensory perception, a sixth sense which lies somewhere between the natural and the supernatural, eerie but explicable. This power, it seems, can be acquired through a book learning of sorts (*L'Histoire du cœur humain*, for example). The young traveler-initiate in this tale merely undergoes a colorful, dramatized version of the learning process every student of Marivaux's social science texts should experience: the acquisition of the power to unmask his or her fellow humans through heightened perception.[11]

The analogies that can be made between physiognomical theory in this tale and as it exists in other physiognomical writings we shall study are fairly clear. As well as the definition of physiognomical signs as constituting a language more reliable than that of words, physiognomical theories often center on the necessity for heightened perception on the part of the physiognomist. Throughout his journals, Marivaux consistently focuses on the character of the spectator, who spends his life observing and unmasking others. Marivaux's "spectator" is analogous to the figure of the physiognomical *génie* in Lavater and that of the *observateur du coeur humain* in Balzac. Lavater, for example, will insist that although the perceptive skills of a physiognomist can be honed through acquired knowledge, physiognomy is more a phenomenon of intuition than of intellection. This belief in an emotional, completely reliable ability to interpret physiognomical signs, an ability that renders systematized knowledge unnecessary, is the basis for Marivaux's theory of the spectator as well.[12]

Marivaux's spectators state explicitly that they are not authors but observers. In fact, Marivaux begins his first periodical, *Le Spectateur français*, with a warning to his readers that they are not reading the work of a professional writer:

> Lecteur, je ne veux point vous tromper, et je vous avertis d'avance que ce n'est point un auteur que vous allez lire ici. C'est un homme, à qui dans son loisir, il prend une vague envie de penser sur une ou plusieurs matières; et l'on pourrait appeler cela réfléchir à propos de rien. Ce genre de travail nous a souvent produit d'excellentes choses, j'en conviens; mais pour l'ordinaire, on y sent plus de souplesse d'esprit que de naïveté et de vérité . . . (114)

Once he has thus established the distinction between his spectator (who is basically the eighteenth-century equivalent of the seventeenth-century *honnête homme*) and a more systematic, intellectual, professional writer, Marivaux makes it even more explicit why the observations of the former are more valuable:

> Il y a toujours je ne sais quel goût artificiel dans la liaison des pensées auxquelles on s'excite. Car enfin, le choix de ces pensées est alors purement arbitraire, et c'est là réfléchir en auteur. Ne serait-il pas plus curieux de nous voir penser en hommes? En un mot, l'esprit humain, quand le hasard des objets ou l'occasion l'inspire, ne produirait-il pas des idées plus sensibles et moins étrangères à nous qu'il n'en produit dans cet exercice forcé qu'il se donne en composant? (114)

Marivaux's spectator is thus a thinker who rejects "arbitrary" thought (that is to say, thought which is organized or systematized), in favor of thought which is inspired by the sight of certain objects or events. This spontaneous thought is more natural and therefore more reliable, and can only be defined as intuitive rather than intellectual. This opposition is stated even more clearly later in the same *feuille:* "Je préférais toutes les idées fortuites que le hasard nous a données à celles que la recherche la plus ingénieuse pourrait nous fournir dans le travail" (116–17). Marivaux's spectator is a sort of naïve genius, free of guile and artifice and full of innate wisdom. The figure of the physiognomical "genius," in Marivaux as well as in later avatars such as Lavater's physiognomist and Balzac's *observateur du coeur humain,* is marked by his congenital aptitude for analysis of his neighbor, the intuitive nature of his acute perception, and the simplicity and virtue of his own character.

In a study entitled *Ombres et lumières dans l'œuvre de Pierre Carlet de Chamblain de Marivaux,* Suzanne Muhlemann undertakes an extensive analysis of the opposition of intellection and intuition in Marivaux's work, and of the centrality of intuition in his writing. The following passage gives a brief but complete summary of her very accurate analysis:

> L'intuition nous permet de saisir instantanément une vérité. La connaissance que nous avons de celle-ci est sûre et irrévocable, car l'élément clairement aperçu nous apparaît dans un réseau d'interdépendances qui, bien qu'obscures à notre intellection, constituent en quelque sorte la substance sensible de notre vie et par cela même la base de notre connaissance.
>
> En nous apportant l' "instruction sans clarté," l'intuition nous permet de saisir la vérité, c'est-à-dire, la réalité des êtres et des choses qui nous entourent, instantanément et absolument.[13]

If we can establish such a system of thought (or nonthought) as a foundation of Marivaux's physiognomical observations, it is certainly as prevalent in his works of fiction as in his periodicals. As we saw in "Le Voyageur dans le Nouveau Monde," Marivaux uses this type of naïve yet intuitively brilliant narrator to recount a narrative in the first person. Rather than have his narrator expose any theory or system of social science, he merely has him/her stumble upon significant discoveries about his/her fellow beings and share his/her wisdom with the reader.

La Vie de Marianne

The spectator persona, Marivaux's physiognomist, born in the periodicals, is nowhere more compellingly incarnated than in the title character of *La Vie de Marianne.* Marianne insists throughout the course of her narrative that she is not an author ("je ne suis point auteur") and that she knows neither how to write nor how to express complex abstractions. The fact that she tells her reader this in the midst of a narrative characterized by extreme verbosity and a penchant for philosophical digressions is a significantly ironic point not to be overlooked.[14] She begins her story by complaining to the friend to whom she is writing her memoirs that she has no literary competence:

> Quand je vous ai fait le récit de ma vie, je ne m'attendais pas, ma chère amie, que vous me prieriez de vous la donner toute entière, et d'en faire un livre à imprimer. Il est vrai que l'histoire en est particulière, mais je la gâterai, si je l'écris; car où voulez-vous que je prenne un style?[15]

The above passage should not, however, convince us that Marianne truly believes her lack of verbal style renders her ineffectual as an observer of the human condition. While advertising her naïveté, she, like the primary voices of Marivaux's journals, also believes that it is this very naïveté that validates her observations and makes them all the more genuine:

> Je ne sais point philosopher, et je ne m'en soucie guère, car je crois que cela n'apprend rien qu'à discourir; les gens que j'ai entendu raisonner là-dessus ont bien de l'esprit assurément; mais je crois que sur certaine matière ils ressemblent à ces nouvellistes qui font des nouvelles quand ils n'en ont point, ou qui corrigent celles qu'ils reçoivent quand elles ne leur plaisent pas. Je pense, pour moi, qu'il n'y a que le sentiment qui puisse nous donner des nouvelles un peu sûres de nous, et qu'il ne faut pas trop se fier à celles que notre esprit veut faire à sa guise, car je le crois un grand visionnaire. (60)

The above passage clearly links Marianne with the other spectators in Marivaux's work, and alludes to the fact that she is a confident spectator herself, relying on an instinctive ability to read her fellow beings, as the rest of the novel largely illustrates.[16] Indeed, almost everything that happens to Marianne in the course of the extremely eventful period of her life covered by the novel is analyzed by Marianne in either physiognomical or (even more frequently) pathognomical terms. In any given situation, Marianne, like the narrator of "Le Voyageur dans le Nouveau Monde," seems to have a more than usual ability to assess the real thoughts and feelings of those involved and to distinguish between their interior and exterior selves.[17] In order to come to these often accurate conclusions, Marianne the true physiognomist uses clues given by physical appearance to gain access to true thoughts and character, decomposing signs (people) into signifiers (physical traits) and signifieds (metaphysical traits).

The examples of Marianne's physiognomical skills are far too numerous to be cited in their entirety, but a few instances would perhaps be appropriate here:

[on M. de Climal, the lascivious *faux dévot*] A peine lui eûs-je répondu cela, que je vis dans ses yeux quelque chose de si ardent, que ce fut un coup de lumière pour moi; sur le champ je me dis en moi-même: il se pourrait bien faire que cet homme-là m'aimât comme un amant aime une maîtresse . . . tout d'un coup les regards de M. de Climal me parurent d'une espèce suspecte. (72)

[on the *beau monde* in the church scene in which she first encounters Valville] C'étaient des femmes extrêmement parées . . . J'en vis une fort aimable, et celle-là ne se donnait pas la peine d'être coquette; elle était au-dessus de cela pour plaire; elle s'en fiait négligemment à ses grâces, et c'était ce qui la distinguait des autres, de qui elle semblait dire: je suis naturellement tout ce que ces femmes-là voudraient être . . . Et moi, je devinais la pensée de toutes ces personnes-là sans aucun effort; mon instinct ne voyait rien là qui ne fût de sa connaissance . . . (88)

[in the awkward scene in which M. de Climal finds her with M. de Valville] Tout cela était difficile. Il pâlissait et je ne répondis rien; ses yeux me disaient: Tirez-moi d'affaire; les miens lui disaient: Tirez-m'en vous-même; et notre silence commençait à devenir sensible . . . (111)

[on greeting Valville in the *parloir* of her convent] Qu'il était bien mis, lui, qu'il avait bonne mine! Hélas! qu'il avait l'air tendre et respectueux! Que je lui sentis l'envie de me plaire, et qu'il était flatteur, pour une fille comme Marianne, de voir qu'un homme comme lui mit sa fortune à trouver grâce devant elle! Car ce que je dis là était écrit dans ses yeux . . . (190)

The above examples are but a few of the constant workings of the mind of Marianne the spectator. It is important to note, however, that the above are all examples of her pathognomical perception, and not (to speak in strict Lavaterian terms) of her physiognomical perception. While Marianne is an unusually keen observer of the facial expressions and gestures of her fellow beings, and while the novel therefore contains an inordinate quantity of such information (as well as an inordinate number of plot twists provoked by such observations), this phenomenon is not especially unusual in an eighteenth-century novel.

Of greater interest to the student of physiognomy and of character portrayal in the novel are its much commented set-piece portraits. These literary portraits, descendants of both the *précieux* salon game of portraiture and the seventeenth-century tradition of the *portrait moral*, were standard features of a certain tradition of the eighteenth-century novel.[18] Unlike the ongoing pathognomical observations, which are interpretations of facial expression or gestures in a given situation, usually more or less integrated into the narrative, the portraits are lengthy, stylized portrayals of the various aspects (physical, moral, and social) of a character. The portraits depict a character in stasis, in his/her essential, unchanging qualities, rather than in his/her actions and physical or moral evolution. The distinction between this type of character portrayal and examples of Marianne's *aperçus* above is therefore analogous to that between physiognomy and pathognomy.

The portraits are crucial points in the text at which the narration of Marianne's adventures stops and the interpretation of the atemporal essence of a person in his or her physical, moral, and social components takes over. The portraits are distinguished from the rest of the text in formal, easily recognizable ways. Philippe Hamon, in his essay on "clausules," argues that there are certain places in every text ("lieux stratégiques") where textual substructures (such as a portrait) define their own beginnings and ends, where for example the line of demarcation between narration and description is drawn. These "lieux stratégiques" define what are, in effect (according to Hamon, in another essay on metalanguage and literary texts), "des structures à forte autonomie." Hamon is referring to descriptive passages in general and to portraits as a subset of description when he defines these largely autonomous textual segments.[19] As for the metalanguage which introduces these particular textual set pieces, Marivaux generally makes things quite explicit by actually using the terms "portrait" and "peindre" in introducing the passages.[20]

There are two instances of portraits in *La Vie de Marianne* which meet the general description (including length) of a literary portrait: that of Mme. de Miran, Marianne's benefactor, and that of Mme. Dorsin, friend of Mme. de Miran and admirer of Marianne. Before "painting" either of these portraits, Marianne, in characteristic spectator fashion, disavows her ability to give anything but a very sketchy verbal representation of any other human being, in the following oft-cited passage:

> Quand je dis que je vais vous faire le portrait de ces deux dames, j'entends que je vous en donnerai quelques traits. On ne saurait rendre en entier ce que sont les personnes; du moins cela ne me serait pas possible; je connais bien mieux les gens avec qui je vis que je ne les définirais . . . (169–70)

Lest we too eagerly attribute a poststructuralist skepticism about the representational power of language to Marianne, we must note that

Marianne's remarks serve as a preface to two portraits in which she very effectively renders at least a good outline of her two subjects. While we as readers (and perhaps Marivaux as author) recognize the validity of Marianne's statement on the impossibility of capturing a person "en entier," Marianne the narrator is merely being *coquette*. She denies her ability to paint a portrait so as more effectively to amaze and dazzle her readers in the next few pages, where she indeed paints what might appear to be complete portraits of each of the two women.

The "complete" portrait includes analyses of the subject's physical, moral, and social qualities, much of this expressed in the extremely subtle psychological refinements and abstract distinctions characteristic of *marivaudage*.[21] The portrait begins with the physical, and this of course is the part of most interest to my study. Marianne begins her portrayal of Mme. de Miran by stating her age ("elle pouvait avoir cinquante ans") and goes on to analyze her physical appearance and, more specifically, its effects on others: "Quoiqu'elle eût été belle femme, elle avait quelque chose de si bon et de si raisonnable dans la physionomie, que cela avait pu nuire à ses charmes, et les empêcher d'être aussi piquants qu'ils auraient dû l'être" (170). While the French word *physionomie* is, strictly speaking, a synonym for *visage*, it is clearly being used here as more of a synonym for *air*; indeed the very next sentence replaces *physionomie* with *air*. "Quand on a l'air si bon, on en paraît moins belle; un air de franchise et de bonté si dominant est tout à fait contraire à la coquetterie . . ." (170). This shift from *physionomie* to *air* (which the *Petit Robert* defines as "apparence expressive") makes it clear that Marianne is referring not merely to the material reality of Mme. de Miran's face, but rather to the face as the nexus of the physical (beauty) and the moral (goodness, reason). From the very first sentences, we see that the portrait is a "lieu privilegié" for this confusion of the physical and the moral, for what is indeed physiognomical observation.[22]

Marianne continues the portrait, not with specific description of Mme. de Miran's body, but rather with observations on the relation between her physical, moral, and social selves and general reflections on the effect of women's physical appearance on each other. The closest Marianne comes to any description of the physical free of confusion with descriptions of the moral and social is the following passage:

> Or, à cette physionomie plus louable que séduisante, à ces yeux qui demandaient plus d'amitié que d'amour, cette chère dame joignait une taille bien faite, et qui aurait été galante, si Mme. de Miran l'avait voulu, mais qui, faute de cela, n'avait jamais que des mouvements naturels et néces

saires, et tels qu'ils pouvaient partir de l'âme du monde de la meilleure foi. (171)

The above paragraph is immediately followed by one which begins "Quant à l'esprit . . . ," and thus ends the first segment of the portrait dealing with the physical self.

The portrait of Mme. Dorsin, which appears somewhat later, follows the same basic schema. Marianne begins by stating that she is about to paint a portrait ("c'est ici que j'ai dit que je ferais le portrait de cette dame"). She begins the portrait with a remark about the age of her subject (in relation to Mme. de Miran: "Mme. Dorsin était beaucoup plus jeune que ma bienfaitrice"), and continues from there with comments on the physical person. It is noteworthy that, as in the portrait of Mme. de Miran, the second sentence of the portrait proper (the one which, in each case, follows the remark on age) uses the word *physionomie* to refer to the appearance of the subject. Indeed, in the portrait of Mme. Dorsin, it is Marianne herself who calls attention to this word and to the special connotations it has: "Il n'y a guère de physionomie comme la sienne, et jamais aucun visage de femme n'a tant mérité que le sien qu'on se servît de ce terme de physionomie pour le définir et pour exprimer tout ce qu'on en pensait en bien" (206). Marianne makes the distinction here between *visage* and *physionomie* by saying that the term "physionomie" must be earned, that a *physionomie* is more than just the anatomical entity defined by *visage.* A bit further down, she gives us an idea of what constitutes a *physionomie* as opposed to a simple *visage:*

> Je ne parle ici que du visage, tel que vous l'auriez pu voir dans un tableau de Mme. Dorsin.
>
> Ajoutez à présent une âme qui passe à tout moment sur cette physionomie, qui va y peindre tout ce qu'elle sent, qui y répand l'air de tout ce qu'elle est, qui la rend aussi spirituelle, aussi délicate, aussi vive, aussi fière, aussi sérieuse, aussi badine qu'elle l'est tour à tour elle-même; et jugez par là des accidents de force, de grâce, de finesse, et de l'infinité des expressions rapides qu'on voyait sur ce visage. (206)

The link here is again that between *physionomie* (as opposed to *visage*) and *physiognomonie: physionomie* is the face, but specifically in its capacity as organ of expression of the soul. The importance of Mme. Dorsin's face lies not in physical, aesthetic, or anatomical terms, but rather in physiognomical terms. Marianne lauds a face which "deserves" to be called a *physionomie* not only for its pathognomical capacity (the ability to express passing emotions: "l'infinité des expressions rapides qu'on voyait sur ce visage"), but also for its strictly physiognomical capacity (the ability to reflect essential character traits: "qui y répand l'air de tout ce qu'elle est").[23] As in the portrait of Mme. de Miran, Marianne uses the word *air* in her description of a *physionomie* to refer to a sort of general impression conveyed by physical appearance which often comprises the moral as expressed by the physical (one can have "l'air vertueux," for example).[24] Also, as in the portrait of Mme. de Miran, the last passage of the first, physical segment of the portrait ends on an inarguably physiognomical note ("l'infinité des expressions rapides qu'on voyait sur ce visage") and is followed by a seemingly abrupt shift to a description of the subject's moral self ("Parlons maintenant de cette âme, puisque nous y sommes").

Marianne's wording of these transitions would have us believe that a description of the physical naturally engenders a discussion of the moral, and that the transition is a seamless one. It is indeed smooth, simply because her portraits have included the moral from the beginning and there is in fact little transition to be made. In spite of her claim that she has given us a clear, pictorial image of Mme. Dorsin's physical self ("je ne parle ici que du visage, tel que vous l'auriez pu voir dans un tableau de Mme. Dorsin"), Marianne's depiction of the physical has always been intertwined with and, indeed, dominated by, a depiction of the moral (and, as Peter Brooks points out, of the social).[25] While it would seem logical to assume that a portrait would include at least some basic references to the purely physical appearance of its subject, neither Mme. de Miran nor Mme. Dorsin is described in any but the most vague physical terms. At each moment where a physical description seems imminent, a physical (or even aesthetic, as in the case of *belle*) term is counterbalanced by a comment on the moral; the body is always being read through its relation to the character, and never on its own terms:

> Quoiqu'elle [Mme. de Miran] eût été belle femme, elle avait quelque chose de si bon et de si raisonnable dans la physionomie . . . (170)

> On ne prenait garde qu'elle était belle femme, mais seulement la meilleure femme du monde. (170)

> Mme. Dorsin était belle, encore n'est-ce pas là dire ce qu'elle était. Ce n'aurait pas été la première idée qu'on eût eue d'elle en la voyant: on avait quelque chose de plus pressé à sentir . . . (206)

In other words, each time we might reasonably expect to be given some specifics about the body of the person in question (the comments on beauty would seem a logical introduction), the focus shifts to the metaphysical.[26] "Belle femme" becomes "meilleure femme du monde," and the impression of physical beauty is effaced by the impression of something which is more complex, more conceptual, and less strictly tied to the physical realm ("quelque chose de plus pressé à sentir").

Indeed, the portraits abstain from simple physical description in favor of an analysis which consistently links the physical with the metaphysical; it is the blurring of the distinction between the two domains that links these opening segments of the two set-piece portraits with physiognomical analysis. In *Le Portrait chez Marivaux: étude d'un type de segment textuel*, Hendrik Kars says that the term *portrait*, in Marivaux's work, usually refers to a portrayal of both the physical and the moral:

> On constate une supériorité numérique tres nette des emplois du terme "portrait" se référant à des passages décrivant les deux aspects [physical and moral] à la fois; ensuite, les emplois de la troisième catégorie (aspect moral seul) sont plus nombreux que ceux de la deuxième catégorie (aspect physique seul). En outre, certains emplois de cette dernière catégorie concernent des passages qui, à travers la description du physique, montrent plutôt des qualités morales. . . .[27]

If, on the level of content, the portraits (or at least significant portions of them) can be said to be physiognomical analyses performed by a skilled physiognomist (Marianne), on the level of narrative they are usually seen to be places where (temporal) narrative grinds to a halt in favor of (atemporal) description. This is a widely accepted commonplace of Marivaux criticism. Henri Coulet makes the following comment: "En effet, les portraits ont des défauts: ils interrompent l'action . . ."[28] Vivienne Mylne says: "Another element which halts the progress of the plot is the formal portrait."[29] Kars echoes Coulet: "Or il s'agira en principe de considérer le portrait comme un élément qui interrompt momentanément cette progression pour fixer l'attention sur l'être (l'essence) d'un personnage."[30]

Marianne herself seems to accept the rigid distinction between narrative and analysis, when she introduces her portrait of Mme. de Miran: "revenons à nos dames et à leur portrait. En voici un qui sera un peu étendu, du moins j'en ai peur; et je vous en avertis, afin que vous choisissiez, ou de le passer, ou de le lire" (170). This excessive modesty concerning the interest of her portraits is contradicted, however, by the fact that Marianne "promises" the reader the portraits some time before she actually delivers them, teasing him/her into curiosity about the composition of the set pieces.[31] By the time she is ready to "paint" the first portrait, her comment that the reader can read the portrait or not, as he/she chooses, is merely narratorial coquetry; after having seen the unusual act of charity performed by this most unusually charitable woman, he/she is in all probability eager for some more information about Mme. de Miran. This eagerness is perhaps even more acute for the even more intriguing portrait of Mme. Dorsin (although, significantly, she is a much more peripheral figure to the narrative). Marianne tantalizes her reader with promises for this portrait even more than for that of Mme. de Miran, making him/her wait some thirty pages after the portrait of Mme. de Miran before satisfying his/her curiosity about Mme. Dorsin:

> Il vous revient encore un portrait, celui de la dame avec qui elle était; mais ne craignez rien, je vous en fais grâce pour à présent . . . Je vous le dois pourtant, et vous l'aurez pour l'acquit de mon exactitude. Je vois d'ici où je le placerai dans cette quatrième partie, mais je vous assure que ce ne sera que dans les dernières pages, et peut-être ne serez-vous pas fâchée de l'y trouver. Vous pouvez du moins vous attendre à du singulier. (173)

With such suspense created for what would normally be assumed to be a static textual segment not integrated into the narrative proper, the distinction, at least in this specific case, between narration and description seems to fall apart. Because of Marianne's narratorial coquetry (completely in keeping with the persona of her character), the reader anticipates the portraits as eagerly as the events of the plot, because the two are of equal importance to the integrity of the novel. Because the elements of the plot in *La Vie de Marianne* are invariably inextricable from Marianne's analysis of them (it is this very phenomenon that earned Marivaux his reputation as a tedious nitpicker), it is inappropriate even to attempt to distinguish between the two textual components. It is impossible to read *Marianne* "for the plot," as to a great degree the analysis of events and characters (this includes pathognomical observations, physiognomical observations, and portraits) constitutes the plot. Marcel Arland states this fact nicely: "Il ne suffit pas de dire qu'aux meilleures pages de cette œuvre l'analyse et la vie se complètent, elles se confondent."[32]

Although we have been looking at Marianne as physiognomist up to this point, it is even more suggestive to study her as an object of physiognomy, for it is from this perspective that we will see some of the broader implications of physiognomical thought brought to light in *La Vie de Marianne*. The themes of physical appearance, of essential being, of innate qualities emanating from social class and gender are hardly peripheral to Marianne's story; indeed, they constitute it. Each of these themes is related, directly or indirectly, to physiognomy. Marivaux's novel is filled with physiognomical themes and observations, with the obsessive reading of others that is a legacy of his periodicals.

From the beginning of her story, the theory of the nobility of Marianne's birth, based on no concrete proof (at least during the period of her life covered by the novel), is "proved" by the refined quality of her beauty and manner. As a child, both her physical appearance and her comportment made the theory a credible one to all who knew her:

> J'étais jolie, j'avais l'air fin; vous ne sauriez croire combien tout cela me servait, combien cela rendait noble et délicat l'attendrissement qu'on sentait pour moi. (53)

> Je vous avouerai aussi que j'avais des grâces et de petites façons qui n'étaient point d'un enfant ordinaire; j'avais de la douceur et de la gaieté, le geste fin, l'esprit vif, avec un visage qui promettait une belle physionomie; et ce qu'il promettait, il l'a tenu. (54)

The use, in these key passages, of the overdetermined terms *air* and *physionomie* is significant, for here as in the portraits the analysis is a physiognomical one: Marianne was "jolie," yes, but more importantly she had "l'air fin." The finesse being referred to in this case is not a purely physical one, not the finesse of harmonious, well-defined facial features as opposed to fleshy, ill-defined ones, but rather a refinement of attitude, connoting sensitivity and gentility through the facial features. Jean Molino states the existence of a link between simple physical beauty and the body as legible signifier in Marivaux's work:

> Au thème de la beauté se rattache le thème du corps: nos relations avec autrui et avec nous-mêmes ne prennent pas place au seul niveau du discours, elles font à chaque instant intervenir le corps, dont le langage secret est plus profond, plus primitif, plus vrai, plus efficace que la parole.[33]

Although he does not use the word *physiognomonie*, Molino has essentially described the most fundamental precept of physiognomy, that of the body constituting a semiotic system more natural, less arbitrary, and therefore more reliable than that of language, and defined it as an integral part of Marivaux's thematic system. Nowhere is this precept more central than in *La Vie de Marianne*.

Much is made of Marianne's appearance throughout the course of the novel, and indeed, at many junctures in the plot it is Marianne's pleasing appearance that determines the outcome of a particular situation.[34] This is obviously the case in her relations with both M. de Climal and Valville: they are sexually attracted by Marianne's physical attributes. However, it is perhaps less obvious in the case of Marianne's benefactress, Mme. de Miran. When Mme. de Miran first encounters Marianne, who is weeping in a church, she is filled with pity and compassion, not just for a pitiful young girl, but for a beautiful, refined, pitiful young girl. This episode, perhaps the most important of the novel for determining Marianne's fate, revolves around Marianne's physical appearance and, more specifically, Mme. de Miran's reading of that appearance. Marianne's *physionomie* decides her fate.

As Mme. de Miran enters the church, she hears, then sees the distressed but well-dressed, pitiful but beautiful, young girl:

> A peine y fût-elle, que mes tons gémissants la frappèrent; elle y entendit tout ce que je disais, et m'y vit dans la posture de la personne du monde la plus désolée.
>
> J'étais alors assise, la tête penchée, laissant aller mes bras qui retombaient sur moi, et si absorbée dans mes pensées, que j'en oubliais en quel lieu je me trouvais.
>
> Vous savez que j'étais bien mise; et quoiqu'elle ne me vît pas au visage, il y a je ne sais quoi d'agile et de léger qui est répandu dans une jeune et jolie figure . . . Mon affliction, qui lui parût extrême, la toucha; ma jeunesse, ma bonne façon, peut-être aussi ma parure, l'attendrirent pour moi . . .
>
> Il est bon en pareille occasion de plaire un peu aux yeux, ils vous recommandent au cœur. Etes-vous malheureux et mal vêtu? Ou vous échappez aux meilleurs cœurs du monde, ou ils ne prennent pour vous qu'un intérêt fort tiède . . .
>
> La dame en question m'examina beaucoup, et aurait attendu pour me voir que j'eusse retourné la tête, si on n'était pas venu l'avertir que la prieure l'attendait à son parloir.
>
> Au bruit qu'elle fit en se retirant, je revins à moi; et comme j'entendais marcher, je voulus voir qui c'était; elle s'y attendait, et nos yeux se rencontrèrent.
>
> Je rougis, en la voyant, d'avoir été surprise dans mes lamentations; et malgré la petite confusion que j'en avais, je remarquai pourtant qu'elle était contente de la physionomie que je lui montrais, et que mon affliction la touchait. Tout cela était dans ses regards; ce qui fit que les miens (s'ils lui dirent ce que je sentais) durent lui paraître aussi reconnaissants que timides; car les âmes se répondent.
>
> C'était en marchant qu'elle me regardait; je baissais insensiblement les yeux, et elle sortit. (155–56)

Mme. de Miran looks attentively at the spectacle of the beautiful young girl in tearful but aesthetic abjection, and is predisposed in her favor by her (melo)dramatic attitude of despair ("la posture de la personne du monde la plus désolée"). The moment of truth comes, however, when Marianne's *physionomie* (not her *visage*, it is important to note) is revealed to her. It is Mme. de Miran's pleasure at the sight of this *physionomie* as well as her response to the very theatricality of the scene that seals Marianne's fate, as if (in spite of her favorable impression of her general figure, manner, and shape) she had been waiting to see Marianne's *physionomie* before committing to a full sympathy for her.[35] In this scene, on which the entire plot of the novel hinges, it is clearly Marianne's *body* that determines her fate.

Mme. de Miran's positive evaluation of Marianne's physical appearance is conveyed in a meaningful look, and Marianne responds with a look full of the requisite gratitude and timidity. This is a moment of physiognomical communication par excellence, as the two women exchange the role of spectator and read their futures in each other's faces. Marianne says that "les âmes se répondent" and this may well be true, but what she does not specify is that it is only through the medium of the signifying face, the *physionomie*, that the souls in question are able to communicate. Physiognomical activity is thus not restricted to the parts of the novel that seem to be, according to the traditional distinctions, analysis rather than narrative; it is also an integral part of the narrative. Mme. de Miran's compassion for Marianne is requisite to the sequence of events that make up the rest of the novel, and as we saw in the telling sentence structure of Marianne's account of the scene, Mme. de Miran is *first* pleased by her *physionomie*, and only then touched by her plight ("elle était contente de la physionomie que je lui montrais, et . . . mon affliction la touchait").

The central enigma in *La Vie de Marianne* is that of the uncertainty of Marianne's birth.[36] While it can certainly be unsettling on a personal level not to know the identity of one's biological parents, the problem for Marianne is not merely one of personal or familial identity, but rather the more pressing one of class identity. In the strictly stratified society of *ancien régime* France, the circumstances of a person's life were determined by the class into which he/she was born. A case of ambiguous birth, like Marianne's, is indeed a tragedy in such a culture: since there is no hard evidence that she was the child of the *gens de qualité* traveling in the ill-fated coach, she can not be granted the social status of a *fille de qualité* and must therefore (particularly since she lacks the money necessary to be of the bourgeoisie) be relegated to shop girl status. Of course, the pathos this situation evoked much more successfully among contemporary readers than it probably does today relies on the notion that she *is* in fact the child of aristocrats; if she were the child of their domestic servants, her position as shop girl might even be seen as something of a good deal for her, and the pathos of the novel would be lost.

If we look again at some of the passages cited above, we see that the interest and pity Marianne has the good fortune to provoke among those who meet her are the result not merely of her beauty, but of the refined quality of her beauty and manner. Mme. de Miran is moved by the sight not of a young girl in distress, or even of a beautiful young girl in distress, but of a beautiful young girl whom she assumes to be a *fille de qualité* in distress.[37] Indeed, it is interesting to note that the *parure* which helps win Mme. de Miran's sympathy is a dress especially selected by Marianne for the very specific impression of social class it might give: "je l'avais choisi; il [her costume] était noble et modeste, et tel qu'il aurait pu convenir à une fille de condition qui n'aurait pas eu de bien" (73). Mme. de Miran addresses Marianne as "mademoiselle" (a form of address used only for *filles de qualité*) without hesitation, and when Marianne tells her story, leaving no ambiguity about which of the adults in the coach were her parents ("Je n'avais que deux ans quand ils [her parents] ont été assassinés par des voleurs qui arrêtèrent un carrosse de voiture où ils étaient avec moi; leurs domestiques y périrent aussi . . ." [160]), it is Marianne's appearance and manner that render the story perfectly credible. Indeed, it is important to note that it is not Marianne's somewhat falsified version of her story, but rather her appearance and manner—the "language" of her body itself—that first convince her benefactress of her aristocratic birth. Significantly, the narrative of her life serves to back up the text of her body rather than the more usual and more expected reverse.

Throughout the rest of the novel, it is Mme. de Miran's belief that Marianne has been dealt an injustice that validates her championing of Marianne's cause, for the nobility of Marianne's birth, already apparent in her physical appearance and her version of the narrative of her life, is also demonstrated by her actions. Mme. de Miran makes this quite clear in a defense of Marianne's cause:

> Je suis fâchée qu'elle soit présente, mais vous me forcez de dire que sa figure, qui vous paraît jolie, est en vérité ce qui la distingue le moins; et je puis vous assurer que, par son bon esprit, par les qualités de l'âme, et par la noblesse des procédés, elle est demoiselle autant qu'aucune fille,

de quelque rang qu'elle soit, puisse l'être . . . Et ce que je vous dis là,
elle ne le doit ni à l'usage du monde, ni à l'éducation qu'elle a eue, et
qui a été fort simple: il faut que cela soit dans le sang; et voilà à mon gré
l'essentiel. (297–98)

And, indeed, one might say, it is *l'essentiel* of the novel itself. While
it is true that Mme. de Miran makes what might be interpreted as an
egalitarian statement in the passage above by saying it is Marianne's
virtuous conduct that makes her "demoiselle autant qu'aucune fille,
de quelque rang qu'elle soit . . . ," we must remember that all of the
above is pronounced in the full belief that Marianne is in fact of noble
blood.[38] The comparison, it should be noted, that Mme. de Miran
makes between Marianne and the hypothetical "aucune fille" is not a
comparison between a young girl of completely unknown social origins and one of proven noble origins; it is rather one between a young
girl of certain, but unprovable, nobility and one of proven nobility.
What might appear to be a plea for judging Marianne on her own personal merits, regardless of the facts of her birth, becomes quite clearly
a validation of the theory that her nobility can be proved through both
her face and her actions. Marianne's attributes do not render her a
person worthy of respect in and of themselves; they merely serve as
proof that she is of noble birth and *therefore* worthy of respect. The
distinction in fact between nobility and merit is broken down by the
character of Marianne, as each serves to reinforce the other, if only in
the minds of the other characters.[39]

This is a point on which it is possible to misread *Marianne*, to see it
as a critique of essentialism and a plea for judging people on their personal qualities rather than their hereditary social status. Indeed, the
eventual facts of Marianne's birth remain a blank, so that the reader is
placed in a position of uncertainty vis-à-vis the reasons for Marianne's
"worth." The reader can either join with the overwhelming majority
of characters in the novel who find superficial signifiers (her refined
facial features, her graceful posture) sufficient proof of her aristocratic
origins or maintain a skepticism born of a lack of hard evidence even
while applauding her very capable handling of her various misadventures. That is to say, Marivaux raises the question of the validity of the
criterion of birth as a judgment of innate worth, but because the characters themselves are convinced of Marianne's nobility and because
we know from the title itself that she does, one way or another, obtain
the title of *comtesse*, he never really escapes the notion that she is in all
likelihood of noble birth; this is what makes his questioning safe.[40] Just
as Mme. de Miran, in her defense of Marianne cited above, can allude

to democratic notions because she feels justified in the final analysis by her certainty of Marianne's aristocracy, so Marivaux can allude to these notions throughout the novel while staying safely within the perimeters established by the almost certain, but as yet unproved, facts of Marianne's social identity.

In a much-quoted passage on the question of Marianne's *naissance*, Marcel Arland imagines how the novel might or might not be changed if Marianne were not of noble blood:

> On imagine le bon tour que l'auteur lui eût joué en disant, dans cette douzième partie qu'il n'a pas écrite: Eh bien, non, Marianne n'était pas d'origine illustre; c'etait la fille d'une vachère et d'un bedeau. Et il l'eût dit sans que le caractère ni le destin de son héroïne nous en parussent invraisemblables.[41]

In fact, the scenario Arland describes would completely dismantle the logic of the novel, and could serve only as a pretext for a perverse (albeit amusing) pastiche of the work.[42] The destiny of the heroine, as we discussed above, is totally dependent on the belief on the part of her fellow characters that, as her face proves, she is a misplaced noblewoman: without this belief, Valville might not have fallen in love with her, Mme. de Miran would not have taken such a great interest in her, and numerous minor characters would perhaps not have done their bit to encourage and defend her. Hard as it may be for a twentieth-century reader to imagine and impossible as it is to substantiate, it is nonetheless safe to assume that many (not to say most) eighteenth-century readers probably sympathized with Marianne for precisely the same reason. Marianne's story is fueled at each juncture by her supposed birth, and the supposition is validated by her *physionomie*. In the character of Marianne, noble—and legible—essence is contradicted by uncertain experience, and that contradiction is the very heart of the work.

The notion that Marianne's character is predetermined by her blood and legible through her body is, as is much physiognomical thought, based on the essentialism of heredity. According to the logic of the novel, the essence of Marianne's being is inevitably a noble essence: both her actions and her physical self can only be the result of a genetic predisposition specific to a certain social class.[43] This class essentialism is not, however, the only one at play in the character of Marianne.

Equally important, if perhaps less immediately striking, is the essentialism based on gender which also determines Marianne's fate and directs the course of the narrative. The female body, traditional object of fetishism in Western culture and perhaps at no period in history more so than in eighteenth-century France, is simply more often scrutinized and objectified than the male.[44] As I noted in my discussion of the physiognomical tradition, the anatomical differences which define gender have traditionally been thought to reflect differences of character and essential being between the genders as well. This is what one might call the "degree zero" of physiognomical gender determinism. In *Marianne,* however, the question of gender and its relation to physiognomy and essentialism is much more complex than that facile statement.

Marianne's beauty is linked, as we saw above, with the notion of her nobility; it is nonetheless linked with her gender as well, for traditionally it is the female whose beauty is evaluated and the course of whose life is determined by this evaluation. M. de Climal's response to Marianne's beauty determines the type of "help" he offers her, Valville's response to her beauty leads him to become infatuated with her, all of which eventually leads her in distress to the church where she meets Mme. de Miran, whose response to Marianne's beauty leads her in time to permit her son to marry Marianne. It is thus that the basic framework of the plot of the novel is established with Marianne's body as its center.

More specifically, it is Marianne's vanity about her beauty that reveals the essentialist definition of femininity at work in the novel. For Marivaux, woman is characterized largely by coquetry, vanity, and the manipulation of others. If Marianne's body and manner are the signifiers of the nobility of her blood, her conscious uses of these advantages are the signifiers of the other side of her essence, her femininity. In *Marianne,* vanity is inextricably linked to an essence of femininity, and Marianne is exemplary of the phenomenon:

> Il est vrai que, du côté de la vanité, je menaçais d'être furieusement femme. Un ruban de bon goût, ou un habit galant, quand j'en rencontrais, m'arrêtait tout court, je n'étais plus de sang-froid . . . (82)

> Il me tardait d'aller à l'église pour voir combien on me regarderait . . . nombre de passants me regardèrent beaucoup, et j'en étais plus réjouie que surprise, car je sentais bien que je le méritais; et sérieusement il y avait peu de figures comme la mienne, je plaisais au cœur autant qu'aux yeux, et mon moindre avantage était d'être belle. (84)

Lest we question the fact that this particular type of personal vanity (about the beauty of one's body) is specifically female, Marianne makes explicit the essential quality of the link between gender and character:

> Et moi, je devinais la pensée de toutes ces personnes-là sans aucun effort; mon instinct ne voyait rien là qui ne fût de sa connaissance, et n'en était pas plus délié pour cela; car il ne faut pas s'y méprendre, ni estimer ma pénétration plus qu'elle ne vaut.
>
> Nous avons deux sortes d'esprits, nous autres femmes. Nous avons d'abord le nôtre, qui est celui que nous recevons de la nature, celui qui nous sert à raisonner, suivant le degré qu'il a, qui devient ce qu'il peut, et qui ne sait rien qu'avec le temps.
>
> Et puis nous en avons encore un autre, qui est à part du nôtre, et qui peut se trouver dans les femmes les plus sottes. C'est l'esprit que la vanité de plaire nous donne, et qu'on appelle, autrement dit, la coquetterie.
>
> Oh! celui-là, pour être instruit, n'attend pas le nombre des années: il est fin dès qu'il est venu; dans les choses de son ressort, il a toujours la théorie de ce qu'il voit mettre en pratique. C'est un enfant de l'orgueil qui naît tout élevé, qui manque d'abord d'audace, mais qui n'en pense pas moins. (88)

A more direct articulation of the traditional sexist definition of woman as vain and intuitive, as opposed to man as objective and rational, can hardly be imagined. The essence of woman is tied to intuition, and to the desire to please men and best other women. These supposedly feminine qualities give a woman the ability to read other people (that is to say, other people's bodies) without any recourse to intellection. Marianne, as a woman and therefore necessarily a coquette, does not need to undertake the mystified voyage of initiation described in "Le Voyageur dans le Nouveau Monde." Intuition, the fundamental prerequisite for physiognomical observation, is innate to women, regardless of their powers of reason.

Perhaps most important, we must recognize that this supposed feminine intuition is undeniably linked to the body itself. Women's insights into others, as Marianne explains, come from their own will to please as corporeal objects, from their strong desire to be found beautiful by both men and other women. Particularly intelligent coquettes, like Marianne herself, are characterized by their keen intuitive ability to present the face which will be most appealing to the spectator of the moment. Women thus become expert at reading other people's perceptions of their bodies; which perceptions are in turn themselves legible only because they are registered somewhere on the body of the other person. The kind of intelligence which Marianne defines as specifically and essentially female, then, is one whose foundations are corporeal. The physiognomical "science" of coquetry consists of a

hyperawareness on three levels: that of one's own body, that of potential comparisons of one's own body with other bodies, and that of
others' reactions to one's own body (legible on their bodies). It is for
precisely this kind of remarkably complex corporeal perception that
the character Marianne is famous.

It may be tempting to interpret this as a positive view of woman
and see it, as does Leo Spitzer, as a *"glorification de l'intuition féminine,
organe de connaissance . . ."* and believe that "le sujet de *La Vie de Marianne* n'est pas tant le récit de telle vie de jeune fille intrépide, mais la
glorification *du principe féminin dans la pensée humaine . . ."* [45] However,
it is clear enough from the passage cited above that the novel does not
attribute an ability to *think* to women and that Spitzer's choice of the
word *connaissance* (as opposed to *savoir*) is more suggestive than his
statement about "la pensée humaine." As we saw above in the spectator persona, a man who follows his intuition is natural, guileless,
innocent. A woman's intuition, while providing her with inarguable
perception concerning others, inevitably leads to the artifice and guile
that constitute coquetry.

The point of Marianne's definition of feminine intuition and coquetry is that it has absolutely nothing to do with powers of reason
or intellection or thought.[46] Although women may have certain of
these powers (notably Mme. Dorsin, whose intelligence is described
as "mâle"), Marianne seems somewhat skeptical about their ability to
really develop them ("celui [le pouvoir] qui nous sert à raisonner, suivant le degré qu'il a, qui devient ce qu'il peut . . ."). As described by
Marianne, these powers of reason are personal and, one might even
think, accidental when they occur in a woman. The power of intuition
and its corollary, coquetry, however, are held to be essential to, and
shared by, all women. Furthermore, it is important to note that if the
supposedly masculine qualities of reason and intellection exist theoretically in a purely mental realm, feminine intuition is always and
inescapably linked to the considerably less "pure" realm of the body
itself.

Can a power born of vanity, envy, and petty interpersonal competition truly be considered a "glorification"? Are women exalted by a
view which defines them as being destined from birth, by the simple
fact of their gender, to limited powers of reason but vast powers of corporeal intuition, exaggerated by a continual jealousy and an inevitable
struggle for power? Spitzer himself realizes that such is the nature
of Marivaux's "principe féminin": "Et la vieille Marianne qui écrit ses
aventures se rend bien compte de ce que le caractère congénital de
l'esprit de la femme est à base d'orgeuil . . ." [47] To readers informed by

late-twentieth-century feminist thought, it is clear that Spitzer fails to recognize the fundamental contradiction in his essay, the fact that his accurate, if unintentional, reading of gender essentialism in *Marianne* renders his "glorification" theory inescapably naïve.[48]

It is pertinent at this point to recall the primal scene of spectatorship in Marivaux's work, the first anecdote in *Le Spectateur français* in which the future spectator surprises his mistress before her mirror, practicing her "natural" gestures. It is this shock that disabuses the young man of his social naïveté and, more specifically, reveals to him that behind every woman lies a coquette. It would perhaps not be an exaggeration to say that in his statement "c'est de cette aventure que naquît en moi cette misanthropie qui ne m'a point quitté . . ." (118), the word *misanthropie* could well be replaced by *misogynie*.[49] While it is certainly true that the journals contain many very critical portrayals of men, it is equally true that these men are invariably being critiqued for certain specific individual character flaws: those of vanity and the will to personal superiority. These flaws, according to *Marianne*, are essential to woman. We could therefore make the logical assumption that, just as reason is accidental in woman and essential in man, coquetry (as defined by the traits named above) is essential to woman and accidental in man.[50] If we can say that reason as the product of intelligence is an admirable quality and coquetry as the product of excessive pride and insecurity is a somewhat less than admirable quality,[51] it is clear that the essentialist "principe féminin" as defined by Marivaux in *Marianne* is anything but glorified, regardless of what the author's intentions may have been.[52]

In a novel obsessed with the human body, with beauty, and with both physical appearance itself and its consequences, there is surprisingly no concrete description of physical appearances. It seems curious that for all the importance accorded to Marianne's beauty and its signifying powers, there is virtually no physical detail given about her.[53] A reader might well finish the novel with a very concrete image of Marianne's appearance, only to realize that the details of this image were in no way supplied by the novel itself.[54] Although we know quite well that Marianne has beautiful hair, are we sure what color it is? We know that she possesses a "petit minois" and that it has cost her "pas mal de folies," but is her nose aquiline or *retroussé*? Are her extremely perceptive eyes, principal actors in so many scenes of intense pathognomical exchange, blue or green or brown? Is she tall and slim or short and "petite"?

The lack of specific physical detail about the protagonist might be explained by the fact that *Marianne* is, after all, a novel written in the first person. However, it is hardly in keeping with the persona of the narrator to keep us in the dark about information which might reflect well on her (she certainly doesn't fail to tell us that she is beautiful, only to provide us with the specifics of her beauty). Even more pertinent is the fact that she gives no more physical detail about other characters than about herself.[55] There is certainly ample opportunity within the perimeters of the portraits to give a few descriptive details about the physical appearances of her subjects, and it would seem logical for her to mention something concrete about the physical person of Valville, with whose appearance she becomes so abruptly infatuated.

Although this lack of corporeal description may seem odd, it makes sense if we realize that *La Vie de Marianne* is not so much a novel about the body as a novel about the effects of the body on those who are confronted with it. The color of Marianne's eyes is of no import to the internal coherence of the novel, but the effects they have on others is of fundamental import to the narrative. The various responses to Marianne's appearance provide the framework for the narrative, ranging from lust (M. de Climal) to infatuation (Valville) to sympathy (Mme. de Miran), and it is therefore these responses Marivaux emphasizes.

This important distinction notwithstanding, the imprecision of some of the terms used to characterize Marianne is remarkable. None is more telling than the word *aimable*. Marianne is frequently described, by both herself and others, as being *aimable*. The following are but a few examples:

> Elle est charmante, et . . . en vérité je ne sache point de figure plus aimable, ni d'un air plus noble. (174)

> Et qui m'a dit de bonne foi . . . que la jeune enfant était fort aimable, qu'elle avait l'air d'une fille de très bonne famille . . . (178)

> Linge assez blanc, mais toujours flétri, qui ne vous pare point quand vous êtes aimable, et qui vous dépare un peu quand vous ne l'êtes point. (189)

> [after Mme. de Miran consents to Marianne's marriage to her son] Pourquoi pleures-tu? . . . je n'ai rien à te reprocher; je ne saurais te savoir mauvais gré d'être aimable . . . (196)

> J'entendis même que ce jeune homme disait à l'autre du ton d'un homme qui admire: Avez-vous jamais rien vu de si aimable? (286)

> Qu'elle est aimable! Nous n'avons rien de si joli à Paris. (306)

In most of the examples above, *jolie* might seem a suitable synonym for *aimable*. However, unlike *jolie*, *aimable* has an inherent etymological ambiguity, the primary meaning of *aimable* being of course, according to the *Petit Robert* (1985), "qui mérite d'être aimé." Furetière's *Dictionnaire universel* (1690) gives the following definition: "Qui a des qualitéz qui attirent l'amour, ou l'amitié de quelqu'un. Cet homme est fort *aimable* par sa belle humeur. Cette femme est *aimable* par sa beaute . . ." [56] The *Dictionnaire historique de l'Academie francaise* (1858–88) tells us that "*Aimable*, avec le temps, a perdu de sa force primitive; au dix-huitième siècle et depuis, il s'est dit particulièrement, dans le langage de la société, des personnes qui plaisent par leurs agréments." When we research the word *agréments* we find it defined by the same *Dictionnaire historique* as "des qualités par lesquelles plaît une chose ou une personne . . ." For its part, the *Petit Robert* says: "Qualité d'une chose, d'un être, qui les rend agréables . . ."

The tautology of these definitions does nothing to dispel the basic ambiguity of the terms (at least for our purposes): does *aimable* (and by extension *agréments* may be included in the question) refer to pleasing physical features (in which case *jolie* could indeed be substituted) or to metaphysical traits (which are normally considered to be the appropriate objects of love)? This is a question which can not be answered, for even if we were to find a definition in a dictionary more or less contemporary to *Marianne* which made explicit that the word could refer to strictly physical beauty, the word would retain an ambiguous connotation by virtue of its etymology.[57] It is unimportant whether the usage was a common one at the time or specific to Marivaux; what is important for us is that in *Marianne*, it is a word frequently used to describe the protagonist which inevitably confuses the physical and the spiritual. When Marianne is said to be *aimable*, is she being called pretty or lovable? Or perhaps, lovable because she is pretty?[58] To a reader who is sensitive to such distinctions, the word *aimable* as it is used to refer to Marianne is both maddening and highly suggestive in its ambiguity.

The ambiguity of the word *aimable* can of course be said to be the ambiguity that constitutes physiognomy itself, in which the physical and the metaphysical are always either inextricable from each other or in a sort of cause-and-effect relationship. Indeed, as we saw above, the physical in its own right is not given any consideration in the novel; in a novel centered on the theme of beauty, physical description is eschewed in favor of physiognomical description. In *Marianne* the responses to beauty, as foundations of the narrative, are more important than the beauty which provokes them. Furthermore, Marianne's beauty is not one which is considered for its own merit, as simply an aesthetic or even sexual object; it is one which must be read and decoded to be understood. Marianne functions as a human sign who must be broken down into the component parts of signifier (her body, and specifically her beauty) and signified (the notions of her noble blood and "femininity"). It is the various propensities of other characters to interpret and respond to this sign that drive the plot of the novel. However, as the novel lacks a conclusion and thus an answer to the central question it poses about Marianne's class identity, we never know whether class is indeed legible through the body. The reader is left to continue speculating as to whether the readings of Marianne's body performed by other characters were valid, and thus by extension, whether the legibility of bodies in general is a viable notion. The older Marianne, narrator of the novel, obviously knows the answer to the question, but does not disclose it to the reader of the novel, whose desire for closure on the levels of both the plot and the larger implications of the novel must remain unsatisfied. The text, by maintaining its reader's ignorance, ensures his/her ultimate passivity.

It is clear that the physiognomical concerns in Marivaux's works are of two kinds: those necessitated by the insincerity of human beings as social creatures (as in "Le Voyageur dans le Nouveau Monde"), and those necessitated by any exception in the rigid class system of proven social identity (Marianne's situation being a case of just such an exception). Marivaldian physiognomy is, in either case, of a slippery, potentially ambiguous sort: one can never be sure one is not being deceived by the object of interpretation, as there is, after all, always the possibility that Marianne is merely an especially clever *coquette* who has learned to mimic the *air* and manners of aristocrats. With emphasis thus implicitly placed on *air*, expression, and human will as well as the fallible reception of physical appearance by would-be physiognomists in the novel, the import of Marivaux's physiognomy is above all psychological. Pathognomy, the study of facial expression which is the legacy of seventeenth-century physiognomical thought, is a social, even "worldly," activity. As such it is firmly grounded in a social and psychological context, and easily plotted into a narrative such as that of *La Vie de Marianne*.

By contrast, Lavater's physiognomy will signal, some thirty years later, a return to earlier physiognomical notions, which privilege morphological analysis rather than *air* or expression, and a belief that the body is most meaningful when read as an immobile, atemporal, unambiguous signifier.[59] Lavater indeed privileges physiognomy over pathognomy, as we shall see in the following chapter.[60] In order to be more certain about his conclusions, he valorizes those features which lie beyond human power to disguise or deceive. So as to avoid having to analyze precisely the kind of indeterminate signifier Marivaux relishes for its psychological complexity, Lavater attempts to effect a return to a physiognomy which predates Le Brun and Descartes's emphasis on the voluntary and mobile features. As we shall see, Lavater's work is predicated on a desire to reinscribe physiognomy as a discourse of the absolute, and to establish thereby (once and for all) a physiognomical "alphabet" which transcends all human will, deception, and ambiguity.

Chapter Three

Lavater and *l'alphabet divin*

De quelque côté qu'on envisage l'homme, il est un sujet d'étude; on peut considérer en lui chaque espèce de vie (animale, intellectuelle, morale) prise séparément, mais jamais on ne pourra le connaître que par des manifestations extérieures, par le corps, par sa surface. Quelque spirituel, quelque immatériel, que soit son principe intérieur, quelque élevé qu'il soit par sa nature au dessus de la portée des sens, il devient néanmoins visible et perceptible par sa correspondance avec le corps où il réside . . .

—Lavater

It can be stated with some assurance that the name "Johann Caspar Lavater" was one known to most literate Europeans of the period 1775–1875. It can be stated with equal assurance that it is unknown to most literate Europeans and Americans of the twentieth century. The very fact of his passing from notoriety and widespread name recognition into almost total obscurity is enough to render Lavater an intriguing object of study for cultural history. If we accept the relegation of such anachronistic thinkers as Lavater to the cultural refuse heap, do we not risk the rejection of potentially suggestive tools of cultural and literary analysis? While it would be indeed difficult to argue for the worth of Lavater's work on its own terms (i.e., as a system of determining human character from physical features), it can be read as an idiosyncratic, intriguing, amusing text which attempts to crystallize and systematize physiognomical thought, and is thereby paradigmatic of the rhetorical fallacies and contradictions of all such thought.

Before proceeding to an analysis of Lavater's work itself, however, it would perhaps be pertinent to give a bit of the biographical context of this rather obscure figure.[1] Johann Caspar Lavater was born in 1741 in Zürich. He committed himself early on to a life as a Protestant pastor. He also established himself fairly early as something of a polemicist, embracing such controversial stances as a strictly "Christocentric" Christianity (a theological stance not irrelevant to physiog-

nomy, as we shall discuss below), and the movement for Swiss nation-
alism.[2] In the years that followed, Lavater gained renown as a writer,
preacher, and personality. His version of Christianity was a rather
eclectic mix of mysticism and empiricism, characterized by a desire to
"prove" religious beliefs by natural (and supernatural) phenomena.[3]
His particular theology elicited great controversy and various attacks
in German-language journals. He was especially criticized for alleged
Jesuit sympathies.[4] In spite of his widespread notoriety as a theologian, it was clearly his physiognomical writings that earned Lavater
the celebrity which endured for approximately a century. The last few
years of his life were largely involved with the cause of Swiss national-
ism and resistance against the Jacobins. He died in 1801, after having
been wounded fighting for the cause.

As interesting as it is, the story of Lavater's life is not nearly so interesting as the story of his physiognomical work. Aside from the obviously curious and suggestive content of the work, the facts surrounding its publication, reception, and form are noteworthy.[5] It seems that
Lavater first began planning to write his own systematization of physiognomy in 1772. From the beginning, he conceived of his work as an
exhaustive treatise, a definitive "how-to" manual of the art of reading
character traits from physical appearance. The writing of the opus, entitled *Physiognomische Fragmente zur Beförderung der Menschenerkenntnis
und Menschenliebe*, was a collaborative effort, orchestrated and written in large part by Lavater himself, but containing articles and contributions from a number of his correspondents and acquaintances.[6]
Indeed, it seems that these contributions were included in the final
version of Lavater's work without being modified, and without their
authors' being specifically credited.[7] Among the more famous collabo-
rators were J. G. Zimmermann, a renowned physician and thinker,
Goethe (who supplied both articles and artwork), and Herder.[8] The
original German title of the work, *Fragmente*, is literally true, and in
fact applies to both the content and the form of the work, as we shall
see below.

In his exhaustive study on Lavater, Oliver Guinaudeau tells us that
Lavater's original intention was to write two books, one more "scientific" (it is rather hard to imagine what Lavater might have had in
mind here) and one more tailored to a general audience, with a special
emphasis on the moral implications of physiognomy.[9] At some point,
however, the more technical aspect seems to have been deemphasized
in favor of a popular appeal and a definite theological agenda. The
original German edition was published in four volumes, in Leipzig,
between 1775 and 1778. At the time of negotiating the publication

of this enormous *œuvre*, Lavater seems to have already had an eye
to the international European audience and to have been planning
for both French and English translations. Indeed, in 1774, before the
appearance of the original German edition, Lavater was looking for
translators for the French edition.[10] What is most important to know,
however, is that Lavater himself was an active participant in the preparation of both the French and the English translations, and that these
texts are as "primary" to the student of Lavater as the first German edition. Indeed, the French translation is a reedited and enlarged version
of the German text, and, if we are to accept the opinion of Thomas Holcroft, the English translator of the work, is "a great improvement."[11]
For this reason, as well as for literary-historical concerns pertinent to
this study, the edition I will be specifically concerned with is the original French edition *Essai sur la physiognomonie, destiné à faire connoître
l'homme et à le faire aimer* (The Hague, 1781–1803).[12] It is noteworthy
that both Holcroft's translation (*Essays on Physiognomy; for the Promotion of the Knowledge and Love of Mankind* [London, 1789–93]) and the
contemporaneous Henry Hunter translation (London, 1789–98) are
translations, not from the German, but rather from this original French
edition.[13] Holcroft's translation itself contains an odd sort of affida-
vit from Lavater endorsing its content, adding further legitimacy to a
view of all of these pre-1800 editions as equally primary sources.

Essai sur la physiognomonie

Perhaps the most appropriate place to start an analysis of a work of
physiognomy would be with the physiognomy of the work itself: it is
worth noting that all of the various editions of Lavater were beautifully
presented in large, handsomely bound volumes filled with engravings
of original works by Goethe, Henry Fuseli, the German painter and
illustrator Chodowiecki, and William Blake (who contributed three
plates to the Henry Hunter translation, as well as the frontispiece for
a 1788 English edition of Lavater's *Aphorisms on Man*), among others.
The volumes also contain many striking reproductions of drawings
by such diverse artists as da Vinci, Rembrandt, Raphael, Rubens, Van
Dyke, Poussin, Le Brun, and Hogarth. In fact, in a very significant
way, Lavater's work could be said to be art criticism as well as physiognomy, for he concerns himself as much with these representations
of the human form as representations as with their intended referents.
A characterization of Lavater's volumes as the glossy "coffee-table
books" of their time is neither inaccurate nor irrelevant. Surely the
huge commercial success of the books was due in some part to their
seductive physical appearance as well as to their seductive content.

What exactly is the content of these beautiful tomes? Lavater's physiognomy was certainly the most extensive attempt ever to define and systematize a science of determining human character traits and tendencies from physical appearance. The project was also the most ambitious effort to date to establish physiognomy as a legitimate discipline, and by far the most widely read, before or since.

The title of the French edition tells us right away that the work has a very definite agenda, and is indeed "destiné" to accomplish two goals: the first is clearly in keeping with the agenda of the pseudo-Aristotelian treatise: to "faire connoître l'homme"—that is to say, to impart knowledge, science, truth. As Lavater states it in his preface to the *Essai sur la physiognomonie:*

> Préparé à tout ce qu'on peut attendre du préjugé et des passions, je soutiendrai [les] assauts avec calme et fermeté, convaincu que j'aime et cherche la *vérité* et j'ose ajouter, que je l'ai souvent trouvée—mais pour en convenir avec moi, il faut aussi aimer et rechercher la vérité . . .
>
> On pourrait attaquer ce que je vais en dire, sans qu'elle cessât pour cela d'être une science vraie en elle-même, fondée dans la nature.
>
> Celui qui après avoir lu mon ouvrage contesterait encore cette dernière proposition, douterait ou affecterait de douter de tout ce qu'il n'aurait pas inventé lui-même. (1. v–vi, préface)

The second goal of the work distinguishes it from earlier works on the subject, however, as well as helping to explain its widespread appeal to contemporary audiences.[14] As stated in the French title, this goal is "et à le [l'homme] faire aimer." Lavater makes it quite clear here that he is propagating not knowledge for its own sake, but rather knowledge to be used to a very specific end. The *Essai* is "science" with a moral, theological agenda.[15] Throughout the course of the text, Lavater constantly reminds us that it is so as to better love our fellow man that we should learn to evaluate his appearance. The contradiction embodied in the notion that one must attempt to objectify, to render lifeless through classificatory and generalizing language, in order to love "l'homme" (as such) seems to have completely escaped Lavater. Even more difficult to understand is how he could have failed to recognize the fact that many of the physiognomical judgments contained within his own work can hardly be said to promote love for one's fellow man. Perhaps it is due to the jaundiced perspective of our twentieth-century cynicism, but it is unlikely that many present-day readers could follow the logic that tells us that recognizing our neighbor's sloth and stupidity from his large ears and bulbous nose will lead us to a greater ability to love him. These seemingly apparent contradictions did not occur to Lavater, however, and he clearly considered physiognomy to be a subset of a theology promoting love of one's neighbor.[16]

In no case is the promotion of love for one's neighbor more ambiguous than when the neighbor is a woman. In keeping with ancient physiognomical tradition, Lavater uses the term "fellow man" quite literally: physiognomical discourse is written by men, for men, and (generally) about men. Women are marginalized and dangerous objects of study, to be approached infrequently and with great caution. Lavater devotes a single chapter (or "fragment") to the study of the "sexe féminin," which he prefaces with the following statement of trepidation: "Je commencerai d'abord par avouer que mes observations sur cette moitié du genre humain seront très circonscrites . . ." (4:85). He assures us, with a curious sense of pride, that in fact his knowledge of women is quite limited: "J'ai très peu suivi les femmes dans les occasions où elles peuvent être étudiées et connues . . . Je les fuyais même dans ma première jeunesse, et je n'ai jamais été—*amoureux.*"

Furthermore, in a characteristic moment of paternalism, he lets us know that physiognomy may perhaps be less than kind to women: "J'ai souvent frémi, et je frémis encore, en considérant jusqu'à quel point la Physiognomonie peut compromettre les femmes, à combien d'inconvéniens cette Science peut les exposer." If he is willing to admit that women do indeed have certain positive qualities (patience, indulgence, modesty), and if in fact his readings of individual female physiognomies are much less harsh than his theory, he is nonetheless quick to point out that physiognomy is best used as an arm of defense with regard to the evil of women: "Le vrai sens physionomique à l'égard du sexe féminin est un assaisonnement de la vie, et un préservatif efficace contre l'avilissement" (4:86). If a virtuous woman can indeed ease the pain and fatigue of life for her man, a wicked woman ("ces syrènes impudentes, dont les regards révoltent la modestie et la vertu . . .") can bring about his downfall. Lest we forget that evil is indeed an essential part of the nature of woman (and thus present in even the most virtuous among them), Lavater the theologian reinscribes his characterization of women into the tradition of Judeo-Christian misogyny:

> Elles sont le reflet de l'homme, prises de lui, faite pour lui être soumises . . . Ce n'est pas l'homme qui a été séduit le premier, mais c'est la femme; et l'homme a été séduit à son tour par la femme. (4:89)

Appropriately, this linking of physiognomy and theological tradition would indeed seem to be Lavater's unique "contribution" to the discourse of physiognomical misogyny.

If feminine morality is thus essentially compromised according to Lavater, equally so is feminine intelligence. In an almost eerie re-phrasing of Marivaux's theories of *l'esprit féminin*, Lavater character-izes women as feeling, rather than thinking, beings: "L'homme pense, et la femme sent. La force de l'un consiste dans la réflexion, la force de l'autre dans le sentiment" (4:90). As with Marivaux, it is tempting to emphasize the fact that Lavater assigns a particular "strength" to each of the sexes. However, also as with Marivaux, it is quite clear that the two "strengths" alluded to by Lavater are in fact far from being of equal merit, as we see in the following passage about women: "Irritable par constitution, peu accoutumées à penser, à raisonner et à discuter, entraînées par le torrent du sentiment, elles deviennent aisément fanatiques, et rien ne peut les ramener" (4:90).

This devaluation of the capacity of "feeling" on the part of women is curious in a text which, as we shall see below, ultimately privileges sentiment over intellect. Perhaps even more significant is the ease with which Lavater manages to conflate the physiognomical and the theological to support an essentialist justification for misogyny.

Lavater's discussion of women is however by no means the only, or even the most important, application of theological principles in the *Essai*. As we saw above in my discussion of the subtitle of Lavater's work (". . . à le faire aimer"), theology is the most fundamental under-pinning of Lavater's physiognomy. From the very preface of the *Essai*, Lavater makes plain his theological intent:

> si vous n'apprenez point dans cet Ecrit . . . à mieux connaître et vous-même et vos Semblables et votre commun Créateur à le bénir . . .
>
> Si vous ne sentez naître en vous plus de respect pour la dignité de cette Nature, une douleur plus salutaire de sa dégradation, plus d'amour pour certains hommes en particulier, une vénération plus tendre, une joie plus vive à l'idée de l'Auteur et de l'Original de toute perfection.
>
> Si, dis-je, vous ne retirez aucun de ces avantages—hélas! c'est donc en vain que j'ai écrit . . . (1:vii, préface)

These two rather ambitious goals defined, Lavater continues his preface with a bit of modesty:

> Je ne promets point—car il y aurait de l'extravagance dans cette promesse—de donner en entier l'immense alphabet qui servirait à dé-chiffrer le langage original de la Nature, écrit sur le visage de l'homme et dans tout son extérieur; mais je me flatte au moins d'avoir tracé quelques-uns des caractères de cet alphabet divin et ils seront assez lisibles pour qu'un œil sain puisse les reconnaître partout où il les retrouvera.

> Je déclare ici formellement que je ne veux, ni ne puis écrire un traité
> complet sur la science des physionomies. Je me borne à de simples essais,
> et les *Fragments* que je donne ne sauraient composer un ensemble . . .
> (1:vii, préface)

If the first of the two passages cited here can be read as a relatively simple statement of theological purpose for the work, the latter
is clearly something different. What appears to be a modest disclaimer
on Lavater's part is in fact much more significant: it is an explicit ex-
posure of the rhetorical underpinnings of his system of physiognomy.
There are two rhetorical devices introduced here: the first is a denial
of any pretention to exhaustive mastery of the object of study; the
second is the explicit introduction of the concept of a semiotic system of interpretation based on an analogy with language. Although
the latter of the two will be more obviously germane to an analysis
of physiognomy and its relation to literature, the former is nonetheless indispensable to an understanding of the rhetoric of Lavater's text
itself.

In thinking about the fragmentary form of the *Essai*, one must recall
the fact that the original German title of the work was not *Abhandlung*
or *Aufsatz* or any other German word which might be logically translated into French as *essai*; it was, rather, *Physiognomische Fragmente*.[17]
This very title embodies the primary ambiguity of Lavater's work: on
the one hand, his is a claim to science, to the establishment, once and
for all, of a system of reading physical appearance. His work thereby
seeks to distinguish itself from earlier works on the subject ("la plupart d'entr'eux n'avaient fait que piller Aristote") and posits itself as
an absolutely certain source: "j'ose décider sur nombre de figures et de
traits avec une conviction égale à celle que j'ai de ma propre existence"
(1:13). This is surely a claim to mastery, to absolute knowledge of his
subject, to science as represented in the title by the eminently scientistic word "physiognomische." Lavater is, however, always on the alert
for eventual critics, and is thus careful to claim absolute mastery only
of "nombre de figures et de traits." That is to say, his analyses of certain particular faces are sure, but he claims no general mastery of his
subject. Indeed, Lavater is careful to deny any such pretension, both
in his choice of generic self-classification in the title ("Fragmente")
and in recurring disclaimers throughout the text: "les *Fragments* que je
donne ne sauraient composer un ensemble" (1:vii, préface).

It is fairly clear from the first few pages of Lavater's work that his
is, perhaps by necessity, a defensive project. He makes frequent references to eventual critics ("certains diront que . . .", "je soutiendrai
leurs assauts," etc.) and seems to be continually fending them off

while attempting to win the confidence of the reader. He seems to believe that through his explicit denial of any claim to formal coherence or general applicability of his theories, he can not be held responsible for their inconsistencies and contradictions. Lavater justifies himself through the most primitive of rhetorical devices: denial of responsibility for one's statement.

In spite of his protestations to the contrary, however, his physiognomy, as a self-defined system of interpretation, is necessarily subject to demands of coherence and applicability. The contradiction inherent in the title of the work is that which undermines the notion of physiognomy itself: if it is to be included in the canon of sciences, of generalized systems of knowledge with codified methodologies based on particularized experience(s), it must be able to make the jump from particular observations to general principles. Lavater, by the very nature of his work, implicitly claims the ability to make such a jump. His premise that his is a search for the truth—"convaincu que j'aime et je cherche la *vérité* et j'ose ajouter que je l'ai souvent trouvée . . ." (1:v, préface)—is nothing if not a claim to generality, to absolute and applicable knowledge. The subtitle of the work, "destiné à . . ." ("zur" in German), is in itself a pretension to a goal, to a specified result of the work. If he is going to make such statements of intent and claims to "truth," how can his work not be held accountable for its lack of formal and conceptual coherence?

Furthermore, Lavater is quite explicit about his desire to effect the inclusion of physiognomy within the canon of accepted sciences:

> La physiognomonie peut devenir une science, aussi bien que tout ce qui porte le nom de science. Aussi bien que la physique—car elle appartient à la physique. Aussi bien que la médecine, puisqu'elle en fait partie: que ferait la médecine sans la sémiotique et la sémiotique sans physiognomonie? . . . Aussi bien que les mathématiques, car elle tient aux sciences de calcul, puisqu'elle mesure et détermine les courbes, les grandeurs et leurs rapports, connus et inconnus. (1:62)

Lavater's agenda in writing his physiognomy includes its establishment among the sciences, but he does not succeed (or even really attempt) to justify this establishment on appropriate grounds. Even without specific knowledge about either the historical development or conceptual bases for the sciences he mentions, we can safely presume that each has justified itself through a more or less coherent explication of both its theories and practices, on its own terms.[18] How does Lavater attempt to establish physiognomy among these disciplines? By association with them—because it is in some way related to physics, medicine, and mathematics, it has earned the right to an independent identity as a discipline. Astrology surely has some relation to astronomy, and palmistry to anatomy, but it would be difficult to justify using these relations as reasons for according them similar status as sciences. This is not simply unquestioning adherence to canonical convention; it is an acceptance of the reasons why these distinctions have come to be canonized as conventions. Astrology, unlike astronomy, has to date given neither demonstrable proof of its conceptual premises, nor compelling justification for its practices. The same holds true for palmistry, and, indeed, for physiognomy.[19]

Lavater's definition of his project, as evinced in its very title, is therefore contradictory and self-defeating. He wants to construct a system with a definite agenda and a definite claim to science. On the other hand, he denies any mastery and any formal or rhetorical coherence. The oxymoronic implications of the title *Physiognomische Fragmente* suggest the impossibility of a science of the particular and therefore of physiognomy itself.

It comes as no surprise to discover that Lavater's practice of physiognomy is as contradictory and incoherent as his theoretical and rhetorical agenda. Indeed, Lavater's readings of the individual physiognomies in the *Essai* are unexplained, and any organizing logic his practice may have remains totally obscure to his reader. If we suspend our disbelief for a moment and try to imagine ourselves as earnest seekers of physiognomical method, we come up against the undeniable problem that there simply is no method in Lavater's work. A treatise which posits itself as a "how-to" manual, the *Essai* contains virtually no explication of "how to" whatsoever.[20] The few concrete illustrations from Lavater which follow will suffice, I believe, to demonstrate the uselessness of Lavater's "practical" applications of his science.

Figure 1, which appears in a *fragment* entitled "Exercices physiognomoniques et pathognomiques," is one of many such series of drawings of heads accompanied by brief readings of their purported physiognomical significance. The reading of the very first head is exemplary of Lavaterian practice, beginning with the simple and unexplained statement "Bonté, simplicité, faiblesse." Lavater does not analyze or even identify the features which express in some way these traits; he merely states them as fact, leaving the would-be student of physiognomy to puzzle over the method which led to these very certain conclusions. Although he continues by specifying that the proximity of the nose and the mouth is a mark of imbecility, the usefulness of any such gen-

1. **B**onté, fimplicité, fojbleffe. La proximité du nes & de la bouche
eft une marque d'imbécillité dans les vifages de la forme de celui-ci.
Le derrière de la tête annonce beaucoup de capacité, & ne correfpond
pas au profil.

2. Le haut du vifage a quelque chofe de noble & de fpirituel; le bas eft
dénué d'expreffion.

3. Candeur, bonhommie; caractère pacifique, modefte, fincère, exempt
de paffions — mais foible.

4. Timidité, inquiétude, étourderie, avec une capacité des plus médiocres
& peu de talens.

5. Ce vifage annonce un peu plus d'efprit, & infpire plus de confiance que
le précédent. Ce petit nez camus & cette bouche entr'ouverte ont une
expreffion de timidité; ce grand menton & tout le refte indiquent un
caractère honnête & fans défiance.

6. Le front caractérife un jugement médiocre; l'œil, des paffions nobles &
une forte de grandeur; mais le nez eft commun, & cette bouche de
travers, dont le deffin n'eft pas exact, indique de la foibleffe.

7. Tête manquée d'un homme de génie. La Nature l'avoit bien formée &
bien deffinée; & fi elle n'eft point ce qu'elle devroit être, c'eft aux cir-
conftances qu'il faut s'en prendre: voilà au moins ce que la bouche
femble indiquer. C'eft furtout à l'œil droit, & au fourcil du même
côté, qui eft placé trop bas, qu'on reconnoit que cette tête eft manquée.

Figure 1. Huitième fragment: exercices. Source: Lavater, *Essai sur la physiognomonie,*
2: 72. Yale Collection of German Literature, Beinecke Library.

erality is canceled by the fact that this "rule" apparently holds true
only for faces "de la forme de celui-ci." The problem is of course that
he does not specify what he means by "la forme de celui-ci," and the
reader is forced to guess what renders this face of a particular form as
opposed to another.

The seventh, and last, head of this series is equally indicative of
one of the constant paradoxes of Lavaterian practice. When Lavater
says that this is a "tête manquée," it is not at all clear if he is referring
to the head itself or its representation. In other words, is the head
as it exists in nature not quite that of a genius, or is this a not-quite
accurate rendering of the head of a genius? Given the ambiguity of the

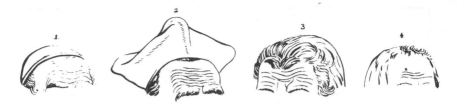

Aen juger par la forme & par les rides, 1 me paroît le plus fage
des quatre. 2 eft plus énergique, plus pénétrant, plus ferme,
mais il eft presque *trop raisonnable*. Le 3 eft un caractère d'airain, qui a
moins de réflexion & plus de force que les précédens. Il ne prend pas
facilement des impreffions, il y réfifte long-temps & s'en défie ; mais une
fois reçues, elles font ineffaçables chez lui. Qu'il fe garde donc bien
d'adopter une idée, à moins d'en avoir fuffifamment reconnu la vérité !
Mon fentiment & mon expérience m'entraînent de préférence vers le 4.
Pureté, générofité, férénité, tranquillité & douceur, il a tout cela, &
en outre un caractère aimant, quoiqu'il mettra dans fes attachemens plus
de conftance que de chaleur.

Figure 2. Addition H. Source: Lavater, *Essai sur la physiognomonie*, 3: 267. Yale Collection
of German Literature, Beinecke Library.

language here ("La Nature l'avait bien formée et bien dessinée . . .";
"c'est aux circonstances qu'il faut s'en prendre . . ."), it is impossible to
answer this question with any certainty. As Lavater's practice is—necessarily, given the nature of his enterprise—as frequently concerned
with critiquing artists' renderings as with reading "nature," the intended object of interpretation (that is to say, the distinction between
referent and representation) is at moments like this almost inevitably
ambiguous.

Figures 2 and 3 are more detailed and would seem to be more analytical. One might indeed expect that they might impart some useful
rules of physiognomical analysis. Figure 2, drawings of four foreheads
(the most significant physiognomical region, according to Lavater),
offers not only an unexplained attribution of general moral qualities
("énergique," "pénétrant," "ferme") to the body parts in question,
but also some completely far-fetched and unsubstantiated speculation
about particular personality traits ("Il ne prend pas facilement des
impressions, il y résiste longtemps et s'en défie . . ."). All of this information is supposedly visible in the form and the wrinkles of the

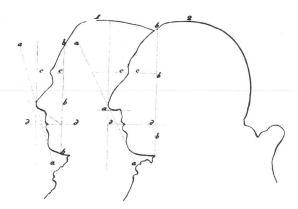

Voici les filhouettes de deux hommes judicieux & pleins de talens, qui malgré l'extrême différence de leurs traits, font liés par une tendre amitié. Les lignes dont nous avons marqué leurs profils, rendent cette différence d'autant plus frappante, & ferviront à la déterminer. Cet exemple nous prouve que la plus parfaite harmonie de fentimens peut fubfifter entre des perfonnes de phyfionomie & de caractère différens, mais non pas hétérogènes. J'accorderois au premier vifage plus de pénétration & de fineffe, au fecond plus de fens & de bonté. A juger de ces deux hommes par le front, je dirois que le premier mène, & que le fecond fe laiffe mener. L'un eft ferme & réfolu; l'autre eft docile & complaifant. Celui-là, délicat fur le point d'honneur, eft emporté par fa vivacité; celui-ci fuit toujours la pente d'un caractère honnête & doux. Le premier doit être en garde contre l'emportement & la précipitation; le fecond doit fe défier de trop de docilité & de lenteur. Pardonnez, couple généreux, fi j'ofe vous juger en public.

Figure 3. Onzième fragment: des silhouettes. Source: Lavater, *Essai sur la physiognomonie*, 2: 182. Yale Collection of German Literature, Beinecke Library.

foreheads, but the reader has absolutely no idea *how*. If, as Lavater tells us, he is guided in some unspecified way by his feeling and experience ("Mon sentiment et mon expérience . . ."), his reader, lacking in all probability this feeling and experience, must fend for himself. The implication is that physiognomical reading is more of an intuitive than an intellectual process, and therefore cannot necessarily be taught. The irony of this notion as the basis for the establishment of a "science" is one which I shall discuss at greater length in further analysis of Lavater's theory below.

Figure 3 is curious in that the silhouettes (a form of physiognomical representation privileged by Lavater) are demarcated by the imposition of a set of geometrical lines. Once again, one is tempted to hope that such iconographic precision implies an equally precise analysis of the semiotics of the profiles in question. Lavater indeed alludes to such an exact analysis ("Les lignes dont nous avons marqué leurs profils, rendent cette différence d'autant plus frappante, et serviront à la déterminer"). However, in keeping with his (surely unintentional) commitment to obfuscation, Lavater does not specifically show the lines to reveal anything, and the reading of the profiles follows no more apparent method than any other in the *Essai* ("J'accorderais au premier visage plus de pénétration et de finesse, au second plus de sens et de bonté").

The fourth and final figure is another silhouette, that of a young boy. Lavater's reading of this profile is exemplary of his lack of justification for the conclusions he so boldly asserts. It is worth noting that Lavater is so certain of the boy's potential that he is willing to stake the entire physiognomical enterprise on the outcome of his predictions (S'il trompe nos esperances—adieu les physionomies").

According to Lavater, even if the boy's "beau front" is hidden by his hair (if it's hidden by his hair, how are we to know that it's handsome?), "we" can, *without supposing anything* ("sans rien supposer"), see immediately that the boy is no ordinary person. The use of the first person plural "nous" (as well as the implied first person plural of "on" in "on ne saurait s'empêcher . . .") is especially curious in this context, as Lavater is obviously seeing something that is much less than apparent to anyone else. Indeed, it is precisely Lavater's failure to include his reader in the process of his very assured physiognomical readings, his inability to create a communal "nous" with respect to the act of interpretation, which renders the "practical" applications of his system as unenlightening as the theory to which they are supposedly linked.

As we can see from the examples of Lavaterian practice above, which are but a few among hundreds in the *Essai*, Lavater's attempts to apply his theories are of little use or interest for his reader. Indeed, it is ultimately as a rhetorical construct, as theory itself, that Lavater's physiognomy is provocative and worthy of analysis for twentieth-century readers.

In reading Lavater, a present-day student of literature is immediately alerted to Lavater's use of several terms which have become virtual "buzzwords" of contemporary literary theory. Lavater defines his physiognomy by analogy with language ("l'immense alphabet qui servirait à déchiffrer le langage original de la nature, écrit sur le visage

J'ajoute la filhouette imparfaite, mais parlante, d'un jeune garçon des plus heureufement organifé. A la nobleffe des fentimens il joint une grande vivacité d'efprit & beaucoup de talens. On ne fauroit s'empêcher de fuppofer à ce profil des yeux d'aigle; mais fans rien fuppofer, & quoique le beau front foit caché par la chevelure, nous voyons auffitôt que ce n'eft pas là un perfonnage ordinaire. Nous fommes obligés d'en attendre de grandes chofes. S'il trompe nos efpérances — adieu les phyfionomies.

Figure 4. Onzième fragment: des silhouettes. Source: Lavater, *Essai sur la physiognomonie*, 2: 183. Yale Collection of German Literature, Beinecke Library.

de l'homme") and with "la sémiotique." To be sure, the eye-catching word *sémiotique* is used here according to its first definition: "partie de la médecine qui étudie les signes des maladies" (*Petit Robert*) and not to the more widely known linguistic and cultural usages later defined by Saussure and Barthes. The use of these concepts (that of deciphering a language and that of semiotics in general) is, however, no mere coincidence; they provide the conceptual and methodological groundwork for Lavater's definition of his "science." Physiognomy, the reading of the body, is of more significance to us as a system of interpretation of signs than as a historical oddity. It is specifically as an attempt to establish a rhetorically coherent semiotics of the body, based on linguistic models, that it is most suggestive, particularly in the context of an inquiry such as mine into the relation of physiognomy and literature.[21]

Lavater has a very specific reason for believing in the necessity of physiognomy: he believes that corporeal, "natural" (i.e., God-given) signs must be interpreted so as to transcend the less trustworthy man-made verbal signs. His goal is to create a method of interpreting bodies which will supersede the necessity of interpreting language itself. Language, Lavater accurately recognizes, is subject to manipulation, falsification, and distortion on the part of an enunciator. Physiognomical signs, on the other hand, reflect an essential truth which their possessor has no power to conceal or transform. Indeed, Lavater tells us, nature itself has provided such reliable signs because it has foreseen the untrustworthiness of human language.[22] Physiognomy purports to provide us with access to a reality beyond that which can be expressed through the manipulation of verbal signs, access to an almost hallucinatory hyperlegibility. It is ironic and significant that the clearest statement of this fundamental principle of physiognomy to be found in Lavater is not his own, but rather a quotation from seventeenth-century French physician and physiognomist Cureau de La Chambre:

> elle [la nature] n'a pas seulement donné à l'Homme la voix et la langue, pour être les interprètes de ses pensées; mais dans la défiance qu'elle a eu qu'il en pouvait abuser, elle a fait encore parler son front et ses yeux, pour les démentir quand elles ne seraient pas fidèles. En un mot elle a répandu toute son âme au dehors, et il n'est point besoin de fenêtre pour voir ses mouvements, ses inclinations, et ses habitudes, puisqu'elles paraissent sur le visage, et qu'elles y sont écrites en caractères si visibles et si manifestes. (1:56)

To borrow the terms of Saussure's semiotic system, Lavater (via Cureau, in the case of the passage cited above) is proposing a reliable set of corporeal signifiers in place of eminently abusable verbal signifiers. This set of signifiers is in fact part of an entirely different language, and it is the interpretation of this better language that Lavater envisages as the result of his *Essai.* Although he admits being unable to supply his readers with this "immense alphabet" "en entier," he is proud to claim having been able to draw "quelques-uns des caractères de cet alphabet divin." Perhaps the most revelatory component of his definition of physiognomy lies in its purpose: it would serve to "déchiffrer *le langage original de la Nature,* écrit sur le visage de l'homme et sur tout son extérieur . . ." (emphasis mine).

The difference between this "langage original" and human language exists not only at the level of the signifier, however; it also refers to the signified. In human language, as defined by Saussure in his *Cours de linguistique générale,* the signified is a "concept," inextricable from and represented by the "image acoustique" that is "signifiant" (signifier).[23] In Lavater's semiotics, not only is this "image acoustique" replaced by the much more material corporeal signifier (large ears, for example), but the "concept" is replaced by an immutable, essential, even allegorical psychological trait (courage, for example).[24] The distinction being made here is clearly that between an original, natural semiotic system and its secondary, inferior reflection, which is human language. Their relationship is that of prototype and imitation. It is precisely because human language is not to be trusted that physiognomy must exist.

Although Saussure serves as an effective point of reference for my reading of Lavater, particularly as a way to *ramener l'inconnu au connu* for late-twentieth-century students of language and literature, it would be inexcusably anachronistic to rely on Saussure as a theoretical context while ignoring extremely pertinent eighteenth-century sign theory. Indeed, many of the concepts of Lavater's semiotic system (and, incidentally, of Saussure's) are echoes of certain pervasive concerns of the thinkers of his time. Lavater could even be read as an attempt to put into practice some of the widely known and accepted theories of language of the period. It is therefore only appropriate for me to give here, as historical context, at least a brief glimpse of these theories.

Perhaps the richest source (indeed, a source to which I will have occasion to return at greater length below) of examples of sign theory in the mid-to-late eighteenth century is to be found in Condillac's *Essai sur l'origine des connaissances humaines* (1746). Condillac defines three categories of signs, including that of arbitrary human language:

> Je distingue trois sortes de signes. 1) Les signes accidentels, ou les objets que quelques circonstances particulières ont liés avec quelques-unes de nos idées, en sorte qu'ils sont propres à les réveiller. 2) Les signes naturels, ou les cris que la nature a établis pour les sentiments de joie, de crainte, de douleur, etc. 3) Les signes d'institution, ou ceux que nous avons nous-mêmes choisis, et qui n'ont qu'un rapport arbitraire avec les idées.[25]

In his *Essai sur l'origine des langues* (circa 1755), Rousseau, for his part, contrasts the arbitrary nature of human language with "natural" systems of communication among members of a single animal species:

> Ceux d'entre eux qui travaillent et vivent en commun, les castors, les fourmis, les abeilles, ont quelque langue naturelle pour s'entrecommuniquer, je n'en fais aucun doute. Il y a même lieu de croire que la langue des castors et celle des fourmis sont dans le geste et parlent seulement aux yeux. Quoiqu'il en soit, par cela même que les unes et les autres de ces langues sont naturelles, elles ne sont pas acquises; les animaux qui les parlent les ont en naissant: il les ont tous, et partout la même; ils n'en changent point, ils n'y font pas le moindre progrès. La langue

de convention n'appartient qu'à l'homme. Voilà pourquoi l'homme fait des progrès, soit en bien, soit en mal, et pourquoi les animaux n'en font point.[26]

Rousseau also makes quite clear his belief, shared by Lavater, in a system of communication through visual signs which supersedes that of human language:

> Ouvrez l'histoire ancienne; vous la trouverez pleine de ces manières d'argumenter aux yeux, et jamais elles ne manquent de produire un effet plus assuré que tous les discours qu'on aurait pu mettre à leur place . . . le langage le plus énergique est celui où le signe a tout dit avant qu'on parle . . .
>   [Ainsi] on parle aux yeux bien mieux qu'aux oreilles. Il n'y a personne qui ne sente la vérité du jugement d'Horace à cet égard. On voit même que les discours les plus éloquens sont ceux où l'on enchâsse le plus d'images; et les sons n'ont jamais plus d'énergie que quand ils font les effets des couleurs. (502–3)[27]

What is perhaps most striking in Lavater is the contradiction between his fundamental distrust of language as a system of signs that can be manipulated and misused and his unshakable faith in a system of signification which functions on precisely the same principles. Lavater sees the problem of language as a superficial one, as one which comes about solely because of its abuse by men of bad faith, not as a problem of the act of signifying itself.[28] The possibility of all fraudulent usage and misinterpretation removed, language would be free of its inadequacies. Lavater therefore envisages a system of signification whose signs are absolute and unquestionable. In so doing, he hopes to escape both the misuse of human language and its inescapably arbitrary nature. Physiognomy is nothing if not a denial of the arbitrary, both in human life and in human language. Lavater makes this intent quite clear in the following passage:

> On ne saurait trop le répéter: attribuer à des causes arbitraires, à un hasard aveugle, sans règle et sans loi, c'est la philosophie des insensés, la mort de la saine physique, de la saine philosophie, et de la saine religion: proscrire cette erreur, l'attaquer par tout où elle se trouve, est l'ouvrage du vrai physicien, du vrai philosophe, du vrai théologien. (1:32)

The distinction between the two languages defined by Lavater is thus the opposition of the essential and the arbitrary. The primary, divine language reflects the absolute truth Lavater seeks (as specified in his preface), an essence of human being. The stasis of such an essence allows it to be defined and known with complete certainty. It is furthermore reflected by the most visible of signifiers, various features of the human body.

In order to prevent misreadings due to manipulation of the features and other forms of corporeal hypocrisy, and to maintain the stasis necessary for his study, Lavater insists on the distinction between his own "science," physiognomy, and pathognomy. Pathognomy, Lavater tells us, is "le miroir des courtisans et des gens du monde" (1:26). It is the study of facial expression, reflections of emotion through the mobile features: the eyes, the mouth, blushing, cringing, etc. Because these are features over which we have more or less control, they are subject to misuse by deceitful people (usually of the worldly variety).[29] The physiognomist avoids the possibility of resultant misreadings by concerning himself with the fixed features, those over which we have no control, and which have therefore been indelibly inscribed with the "langage original" of essential being. Indeed, Lavater suggests that the features at rest, either asleep or (preferably) dead, offer the most reliable opportunities for interpretation. His favorite feature, and the one he finds most suggestive, is thus the forehead.[30]

Indeed, it is Lavater's insistence on defining physiognomy in opposition to pathognomy that renders his work novel. It is also, however, the point which most opponents of physiognomy have found impossible to accept, both before and after Lavater. In the third edition of the *Encyclopédie* (Geneva, 1779), there are two entries for the word *physionomie*. The first tells us that "physionomie" is "l'expression du caractère" and "encore celle du tempérament." Although this correlation between the physical and the metaphysical is offered as the very definition of "physionomie," we are cautioned in the next sentence that "il ne faut jamais juger sur la physionomie" because "il y a tant de traits mêlés sur le visage et le maintien des hommes que cela peut souvent confondre . . ." In a second entry, one which is specifically concerned with *physionomie* as "science imag.," the message is even clearer. The author of the entry begins by telling us that he could well go on himself about "cet art prétendu," but that "M. Buffon a dit tout ce qu'on peut penser de mieux sur cette science ridicule . . ." At this point, a lengthy quote from Buffon's *Histoire naturelle de l'homme* takes over the entry. It is significant that a full century later, in Larousse's *Grand Dictionnaire universel du XIXe siècle* (Paris, 1866–79), a longer and more thorough entry on physiognomy cites the same antiphysiognomical passage from Buffon as its authority in refuting the validity of physiognomy.

Buffon's passage is perhaps the most articulate, and clearly the most widely accepted, refutation of the would-be science. In the part of the *Histoire naturelle de l'homme* entitled "Description de l'homme" (under the heading "De l'Age viril"), Buffon starts out by making the point that man's superiority to all other life forms is clearly illustrated "même à l'extérieur." He goes on in the same paragraph to give us what sounds like an argument for the validity of physiognomy:

> Sa tête regarde le ciel et présente une face auguste sur laquelle est imprimé le caractère de sa dignité; l'image de l'âme y est peinte par la physionomie, l'excellence de sa nature perce à travers les organes matériels et anime d'un feu divin les traits de son visage; son port majestueux, sa démarche ferme et hardie annoncent sa noblesse et son rang.[31]

Further on, he tells us that when the soul is calm ("lorsque l'âme est tranquille"), the features are at rest and "répondent au calme de l'intérieur."

However, when "l'âme est agitée," the features are even more indicative of the soul:

> La face humaine devient un tableau vivant, où les passions sont rendues avec autant de délicatesse que d'énergie, où chaque mouvement de l'âme est exprimé par un trait, chaque action par un caractère, dont l'impression vive et prompte devance la volonté, nous décèle et rend au dehors par des signes pathétiques les images de nos secrètes agitations. (299)

Buffon goes on for several subsequent pages to elaborate on these premises. This is indeed a surprising position coming from a man who is cited as the leading antiphysiognomical authority over a period of a century, from the *Encyclopédie* to Larousse's *Grand Dictionnaire*. It is in fact only after enumerating the powers of signification of various body parts and facial features (with particular attention to the eyes)[32] that Buffon articulates an antiphysiognomical stance. As the transition in Buffon's thought is less than clear, it is probably best to quote him at some length here:[33]

> On peut juger de ce qui se passe à l'intérieur par l'action extérieure, et connaître à l'inspection des changements du visage, la situation actuelle de l'âme; mais comme l'âme n'a point de forme qui puisse être relative à aucune forme matérielle, on ne peut pas la juger par la figure du corps, ou par la forme du visage . . . l'on ne doit pas juger du bon ou du mauvais naturel d'une personne par les traits de son visage; car ces traits

n'ont aucun rapport avec la nature de l'âme, il n'ont aucune analogie sur laquelle on puisse fonder des conjectures raisonnables . . . il est bien évident qu'elles [the "prétendues connaissances" of physiognomists] ne peuvent s'étendre qu'à deviner ordinairement les mouvements de l'âme, par ceux des yeux, du visage et du corps; mais la forme du nez, de la bouche et des autres traits, ne fait pas plus à la forme de l'âme, au naturel de la personne, que la grandeur ou la grosseur des membres fait à la pensée . . . Il faut donc avouer que tout ce que nous ont dit les physiognomistes est destitué de tout fondement, et que rien n'est plus chimérique que les inductions qu'ils ont voulu tirer de leurs prétendues observations métoposcopiques. (304)

In reading the above passage in conjunction with the other passages from Buffon's *Histoire* quoted above, one is immediately struck by the inconsistency of Buffon's view. For example, if he believes that "l'âme n'a point de forme qui puisse être relative à aucune forme matérielle," how can he justify the opening statement of "Description de l'homme" which says plainly, "l'image de l'âme y est peinte par la physionomie, l'excellence de sa nature perce à travers les organes matériels . . ."? These glaring conceptual inconsistencies notwithstanding, Buffon makes an important point in his discussion of physiognomy: the distinction between physiognomy and pathognomy. The basic argument of Buffon's rather confused analysis is that one can read facial expression as an accurate monitor of passing emotions, but that one cannot read fixed bodily form as an expression of moral being.[34] It is precisely Lavater's negation of this opinion that renders his work controversial, novel, and, ultimately, untenable. His choice of the immutable features of the body as the object of his study (as opposed to the more "voluntary" features), of physiognomy over pathognomy, is one which leads us to an analysis of one of the most important aspects of his work.

Lavater's insistence on the distinction between pathognomy and physiognomy (as well as his goal of deciphering an essential language) is motivated by his most basic ideological agenda, the obliteration of the arbitrary.[35] Saussure defines the "signifiant" as being "*immotivé,* c'est-à-dire arbitraire par rapport au signifié, avec lequel il n'a aucune attache naturelle dans la réalité."[36] The opposition between the Saussurean notion of the relationship between signifier and signified, as defined by its arbitrariness, and the Lavaterian language of physiognomy could not be any more distinct. Nowhere is this more dramatic than in their respective uses of the words "nature" and "naturel": for Saussure, the lack of any "attache naturelle" between signifier and signified defines the linguistic sign and consequently language itself;

Lavater's project, on the other hand, attempts to establish the superiority of physiognomical signs by proving that such a "natural"
relation does in fact exist between the "alphabet" inscribed on the
human body and the essential truth it reveals, this act of signification
taking place through the medium of "le langage original de la *Nature*"
(emphasis mine).

It is Lavater's inability to articulate any fundamental difference between the acts of signifying in (arbitrary) human language and in
physiognomy that ultimately renders his claim to the establishment of
an essential language nonsensical. With neither demonstrable proof of
the nature on which he confidently bases his theory nor the rhetorical
force to defend his theory on its own terms, Lavater's physiognomy
is neither a science nor a coherent articulation of a distinct system of
signifying.

The purpose of this study, however, is not simply to prove Lavater
wrong, but also to look at the functioning of his unusual and suggestive system. Perhaps its most suggestive component, particularly
in the context of an inquiry into the relation between physiognomical and literary discourse, lies in the analogies to be drawn between
Lavaterian physiognomy and linguistic and literary models discussed
above.[37] If we look a bit more closely, we find that many more specific
analogies are to be drawn than the primary one of physiognomy as
a system of signifying. For example, the following passage from the
fourth volume of the *Essai* does more than merely reiterate the basic
notion of physiognomy as semiotics:

> A moins de nous faire illusion à nous-mêmes, nous sommes obligés de
> convenir que tout ce qui est *de fait* dans l'homme, se manifeste par des
> *signes*, et que le moindre de ces *signes* répond à un *fait*. Chaque membre,
> chaque particule, chaque muscle, chaque trait, et chaque nuance sont des
> effets de l'ensemble, la fin d'une même origine, le résultat d'une même
> cause. (4:4)

This passage illustrates what we already know about the semiotics
of physiognomy, but also articulates a more specific analogy to be
made with a rhetorical figure, that of synecdoche. Synecdoche is defined by the *OED* as "a figure by which a more comprehensive term
is used for a less comprehensive or vice-versa; as whole for part or
part for whole, genus for species or species for genus, etc." This figure
helps us to define an important component of physiognomy, the belief in an organizing, unifying principle of any human being evinced
through any one of his members. The cruelty of an essentially cruel

man is as legible through his big toe as through his eyes, or as Lavater expresses it: "En un mot il n'est chez l'homme aucune espèce de beauté physique,—ni aucune des parties de son corps, qui ne puisse recevoir de la vertu et du vice (pris dans le sens le plus général) une impression bonne ou mauvaise" (1:149–50).

The figure of synecdoche represents the unquestioning movement between the general and the particular inherent in physiognomy: if we are trying to define a particular character through physiognomy, we express this particularity by attributing general qualities to it. Characterizing a person as "doux, intelligent, et avare" does not particularize him, but merely places him in a series of imaginary constructs, the nebulous categories of those who are "doux," those who are "intelligent," and those who are "avare."[38] The definition of the particular by the general is an invalid one in the domain of science, because it uses an explicitly figurative linguistic device to articulate what Lavater would have us believe is a literal, natural, somehow verifiable relation between body and character. When we say, for example, that someone does not have a roof over his head, we mean of course that he doesn't have walls or a floor either, in short that he doesn't have a house. In such conventional usages or in the context of literary language, the use of synecdoche is valid, as we can safely assume that it is being interpreted for the most part as what it is: a figure of language which refers to some reality only in an indirect fashion. In a text that posits itself as the seminal work of a new science, however, such a premise is hardly justifiable. We need not even subscribe to theories wherein all knowledge is a linguistic construct and all language unreliable; a system of knowledge simply can not be based on a linguistic structure which does not even pretend to any extralinguistic veracity.

An analysis of Lavater's system in terms of linguistic phenomena can be extended to the very agenda of the work. One of the most important goals of his text is to promote the efficacy of physiognomy through the development of a physiognomical vocabulary, a metalanguage comprising the words in human language which best represent physiognomical reality. Lavater recognizes that, in spite of the secondary status of human language as compared with the divine language of physiognomical signs, human language is the tool he must use in order to represent physiognomical concepts.

The goal of a physiognomist must be to interpret, through human language, the nonverbal physiognomical system of signifying. If we break down this moment of interpretation into its semiotic components, we can see that the signified or concept that the physiognomist tries to express is his understanding of the physiognomical sign (itself comprising signifier [body] and signified [essential character]). The physiognomist's signifiers are, however, the words available to him in (human) language. In an attempt to overcome the fundamental inadequacy of language, he must make a conscious effort to strengthen his linguistic power; he must hone his tools of description. Instead of problematizing the fact that the medium of physiognomy is the same human language that it purports to transcend, Lavater insists on the work of the physiognomist as one of working toward a sort of perfect, comprehensive technique of description. He imagines a vast expansion of the physiognomical "alphabet":

> Que n'a-t-on commencé depuis des siècles à étudier la forme humaine, à classifier les traits caractéristiques . . . à commenter chaque fragment; nous aurions à présent l'alphabet du genre humain, alphabet plus volumineux que celui des chinois, et qu'il ne s'agirait plus que de consulter, pour trouver l'explication de chaque visage . . . Je me figure alors une langue si riche, si correcte, que sur une simple description en paroles on pourra retracer une figure . . . (1: 196–97)[39]

In the first half of the passage above, Lavater does not make clear to what kind of "alphabet" he is referring. One could certainly assume that it is the "alphabet" of the transcendent, nonverbal physiognomical language to which the physiognomist seeks access, as outlined in the preface to the work. The last sentence above, however, reveals the unquestioned *glissement* Lavater makes from this rather ambiguous, idealized concept to the more concrete idea of human language ("paroles") as the medium of physiognomy. The language that he imagines as "si riche, si correcte" is none other than ordinary verbal communication itself, elevated to an unprecedented level of descriptive power. The goal of physiognomy here seems to be not the establishment of the existence of an essential, divine language but rather its expression through the medium of a beefed-up (but nonetheless arbitrary) human language.

The notion of the perfection of language as a tool of analysis was, incidentally, a prevalent one in Lavater's time. It is, in fact, one of the tenets of Condillac's method of seeking truth ("la recherche de la vérité") as well as an important concept for the Romantics. Curiously, as if foreseeing Lavater, Condillac specifies that works which treat metaphysical or moral questions necessitate a strengthening of vocabulary and an awareness of the functioning of a system of signs:

> Pourrions-nous jamais réfléchir sur la métaphysique et sur la morale, si nous n'avions inventé des signes pour fixer nos idées, à mesure que

nous avons formé de nouvelles collections? Les mots ne doivent-ils pas être aux idées de toutes les sciences ce que sont les chiffres aux idées de l'arithmétique? Il est vraisemblable que l'ignorance de cette vérité est une des causes de la confusion qui règne dans les ouvrages de métaphysique et de morale. (77)

Ce que j'ai dit sur les opérations de l'âme, sur le langage et sur la méthode prouve qu'on ne peut perfectionner les sciences qu'en travaillant à en rendre le langage plus exact. Ainsi il est démontré que l'origine et le progrès de nos connaissances dépendent entièrement de la manière dont nous nous servons des signes. (220)

As for the physiognomical enterprise itself, human language is indeed the only medium through which the physiognomist can express his physiognomical readings and in fact the only medium through which he can accomplish such conceptualizations. The rather mystical transcendent language to which Lavater refers can be interpreted only through the language whose insufficiency necessitated its existence in the first place. Even if we were to accept Lavater's argument for an essential language of being, legible through the human body, how can we accept a systematization of it which is totally reliant on the inadequate verbal system against which it defines itself?

It is thus somewhat ironic that Lavater's system sets the promotion of human language skills as a goal:

Le physiognomiste ne saurait assez étudier le langage. La plupart de nos erreurs ont leur source dans l'imperfection du langage, dans le défaut de signes parfaitement caractéristiques et adaptés au sujet. Une vérité qui a toute la simplicité et toute la clarté dont elle est susceptible; une vérité rendue avec tous les traits qui lui sont propres, et énoncée avec la précision convenable; une telle vérité ne peut être méconnue de personne . . . Etudiez votre langue maternelle; étudiez les langues étrangères, et surtout la française, qui est si riche en expressions physiognomiques et caractéristiques. Dans vos lectures, dans vos sociétés, vous épierez tous les mots significatifs et vous les noterez dans un vocabulaire. C'est ainsi, par exemple, que vous établirez différentes classes, différentes espèces, pour l'amour, pour le jugement, pour l'esprit, etc. (2: 366)

Lavater tells us that most of our errors in physiognomical judgment come from inadequacies of language, but he does not maintain his original definition of human language as inherently, irreparably inadequate. Instead, he presents the idea that if we perfect our manipulation of language, we can acquire the power to use it to represent physiognomical truths. Significantly, this is one of the very few moments in

Lavater's "how-to" manual of physiognomy where he actually gives his readers practical suggestions for developing physiognomical skills. Truth, he tells us, need only be expressed properly. The expression of this truth necessitates not a new system of signifying, but rather an expansion of one's vocabulary. The problem is no longer that words can not articulate truth or being, but that we often choose the wrong words in our attempt to do so. By thus changing his definition of the problem of physiognomy, Lavater sets a goal for himself and for physiognomists that is both exceedingly more attainable and exceedingly less consistent with the logic of his system.

Fairness, however, compels us to point out that Lavater does acknowledge the limitations of language, in the very significant passage that follows:

> On dit en parlant de plusieurs personnes rassemblées: toutes se sont bien réjouies; et le mot réjouir qu'exprime-t-il, si ce n'est une classe de sensations, différemment modifiées dans chaque individu, et que la situation actuelle modifie de nouveau? . . . Faudra-t-il donc créer un mot, inventer un signe particulier, pour chaque situation individuelle, pour chaque variation, chaque nuance, chaque souffle, chaque mouvement? Ce serait vouloir être Dieu! Ou bien faudra-t-il ne plus parler, parce que tout langage n'est autre chose qu'une classification perpétuelle, et que toute classification est imparfaite et fautive? (1: 105–6)[40]

Here Lavater outlines the undoing of his own system, by articulating the fact that language is by definition a phenomenon of generality. As he says, the variations within the group of possible referents of a given word are so wide as to render the very process of signification through language ludicrous.[41]

The desire to pinpoint physical and metaphysical traits through a widened vocabulary seems fruitless in light of this inescapable inaccuracy of language. At one point, Lavater rightly asks whether our realization of this inaccuracy of language should stop us from talking altogether; our answer is obviously no. First, because not talking is simply not practical, and second, because in most contexts in which language is used, a rough sort of communication is indeed accomplished in spite of the inescapable inadequacies of language.[42] For example, when we ask someone to give us a glass of water, it is probable that he/she will understand and comply with the request as we intended him/her to, in spite of all the variations (both subjective and objective) possible of the referents suggested by the signifiers "give," "glass," and "water." That is to say, language functions in a very practical way when its purpose is to convey an extralinguistic message which is aided by the physical reality of the context (if we ask for the glass of water while standing in a friend's kitchen, he/she can safely infer that "glass" refers to one of those in his/her cupboard and not to some other personal understanding of the concept "glass" we may or may not have). However, the use of language in Lavater's system can not rely on this sort of probable, hit-or-miss act of signifying, simply because his goal is not that of referring to a physical object (like the glass above), but rather that of articulating metaphysical qualities whose only reality is that of the language in which they are expressed.[43] "Virtue," for example, can exist only as we conceptualize it, and we can conceptualize it only through language. Perhaps the oddest of Lavater's notions is the one which holds that it is through painstakingly exact descriptions of the physical that we can gain access to a knowledge of the metaphysical.

If Lavater recognizes that all language is general and incapable of expressing true particularity, how can he set the perfection of descriptive skills as one of the primary goals of the physiognomist? He seems to tell us that the physiognomist must simply do the best that he can, but can this justify attempting to ground a system on an admittedly impossible goal?

Lavater articulates the problem of generality and particularity in language quite nicely in the following passage:

> Qu'est-ce que le langage, en quoi consiste-t-il, si ce n'est en termes, qui expriment des idées générales?
>
> J'excepte les noms propres des hommes, des édifices, des villes, des lieux, et ceux de quelques animaux.
>
> Chaque terme qui exprime une idée générale, est-il autre chose que le nom d'une classe de choses, ou de propriétés, de qualités, qui se ressemblent entre elles, et qui cependant diffèrent aussi à bien des égards? La vertu et le vice forment deux classes d'actions et de dispositions; mais chaque action vertueuse proprement dite, diffère d'une autre action vertueuse, et cette variation est si grande jusqu'au point de séparation où commence le vice, qu'il est souvent telle action qui paraît n'appartenir à aucune classe. (1: 105)

Lavater very justly sees the inevitable generality of language, the fact that words can never really define objects (or people), they can simply classify them. A person who is said to be "beautiful," for example, has been neither described nor defined; he has simply been placed in the category of people whom the enunciator of the word finds beautiful. The acts of defining, describing, and interpreting are in principle acts of particularization, of expressing the particularity of an object through language. As we discussed above, this is necessarily an impossible task, as particularity can not be accurately expressed through a medium of generality. Physiognomy, as defined by Lavater, is the quest for the accurate expression of the particular through the general. It is ironic that the above passage from Lavater serves as a defense against critics of physiognomy who say that "à cause des différences individuelles elle n'admet ni classification, ni abstraction" (1: 105). Lavater uses the requisite generality of language as his defense of the universal nature of physiognomy, but seems to have forgotten that this same generality invalidates the primary goal of his system, the interpretation of that which is most particular.

It is significant that Lavater chooses to except proper names in his definition of the generality of language. A proper name is analogous to the human body in that it is proper (and theoretically particular, although many people may in fact have the same name) to an individual. The proper name is therefore exempt from the generality to which all other words are subject.[44] What Lavater does not mention is that, in order to express that which is particular to an individual, one would indeed need words which are specific to that individual, in the way that a proper name is specific. This is perhaps what he is hinting at when he fantasizes about the rich, innovative vocabulary the physiognomist must develop. The vocabulary of the physiognomist is nonetheless limited to that of the human language(s) he happens to use. Unfortunately, none of these languages has the power to transcend the quality of generality necessary for language to function as an interpersonal means of expression. Lavater's system would work only if he could indeed create a word for each particular variation of each possible human quality, and, as he himself tells us, "ce serait vouloir être Dieu" (1: 105).

Lavater's methodology is, as we have seen above, a semiotic one: a process of reading (physical) signifiers as a means of access to (metaphysical) signifieds. He tells us that man can be known if only one has the requisite interpretative skills which allow the leap from observation of the body to knowledge of the interior being:

> De quelque côté qu'on envisage l'homme, il est un sujet d'étude; on peut considérer en lui chaque espèce de vie (animale, intellectuelle, morale) prise séparément, mais jamais on ne pourra le connaître que par des manifestations extérieures, par le corps, par sa surface. Quelque spirituel, quelque immatériel, que soit son principe intérieur, quelque élevé qu'il soit par sa nature au dessus de la portée des sens, il devient néanmoins

visible et perceptible par sa correspondance avec le corps où il réside, où il se meut comme dans son élément. Ce principe devient aussi un sujet d'observation et tout ce qui dans l'homme est susceptible d'être connu ne peut l'être qu'au moyen des sens. (1: 16)[45]

Perhaps the weakest and most striking element of Lavater's work is dramatically illustrated in the above passage: an absolute belief in the correspondence between the material and the immaterial, without any substantiation of any kind for this belief. The "néanmoins" of the clause which begins "il devient néanmoins . . ." requires some explication in order to be at all credible, but Lavater offers none, neither here nor at any other point in his text. Even more intriguing than his failure to articulate either a theory of this correspondence or a specific hermeneutics allowing such a correspondence to be interpreted, however, is his justification of such an amazing lack of coherence in what is ostensibly a sort of "how-to" manual. If we read the above passage carefully, we see that the only person mentioned is "l'homme" as a "sujet d'étude." Lavater tells us that he becomes "visible et perceptible," but he fails to inform us not only how this is possible, but who is seeing and perceiving. It is indeed Lavater's definition of the person who perceives which, according to his system, justifies his lack of method.

A bit further on in the first volume of the *Essai*, Lavater defines physiognomy in the following way: "J'appelle physiognomonie le talent de connaître l'intérieur de l'homme par son extérieur—d'apercevoir par certains indices naturels ce qui ne frappe pas immédiatement les sens" (1: 22). This definition is a particularly revelatory one in that here for the first time Lavater does not define physiognomy as a "science," an "étude," or even an "art"; he defines it as a "talent." In so doing, he posits the observer or reader as the center of physiognomy. By this definition, physiognomy functions by virtue of a favorable predisposition to the interpretation of signs on the part of the physiognomist. As a talent, unlike a science or field of study, physiognomy cannot be undertaken by just anyone with the will to learn; it is not necessarily something that can be taught. This view obviously puts Lavater's project itself into question: of what use is a manual which proposes to teach that which cannot be taught? A talent is primarily a divine gift, and those who have received the gift of physiognomy, the ability to read appearances, are as immutably and essentially predisposed to this activity as are the objects of their study to their respective characters. The essentialism of physiognomy thus lies not only in the content of physiognomical readings, but in the act of reading itself. As the signs of Lavater's system (made up of natural corporeal signifiers and essential metaphysical signifieds in a natural relation) are predetermined by nature (God), so is the reader of these signs (the physiognomist). The logical conclusion is that the predetermined signs interpreted by the predetermined reader constitute a predetermined reading. Once again, nothing is arbitrary, and physiognomy becomes, in this definition, not only a privileged moment but a sacred one. It affirms itself, Lavater would have us believe, as part of a greater theological *ethos*.

This definition of physiognomy as a talent, as an innate gift, clearly explains the egregious lack of any method or explanation on Lavater's part: as a talent can not be taught, it also need not be taught. One either is or is not a physiognomist. The *Essai* serves to hone the skills of physiognomists through the readings it contains and to awe non-physiognomists with the mystical process. It does not, however, teach anyone physiognomy. Throughout the course of the work, Lavater frequently returns to his definition of physiognomy as being located within the physiognomist, and in fact articulates a cult of physiognomical "genius" within the text. Physiognomical skill is so explicitly defined as an innate ability that one wonders why the *Essai* should have been written at all.

In an analysis of Lavater's system, Michael Shortland tells us that Lavater's project is to be seen as "the establishment of a *science* of *physiognomical perception*." [46] Shortland holds that Lavater defines physiognomy in opposition to pathognomy, scientific physiognomy in opposition to philosophical physiognomy, and physiognomical perception in opposition to physiognomical sensation. The physiognomy/pathognomy distinction is indeed one of the bases of Lavater's self-definition, as we saw above, and Shortland is right to note its significance. However, the other two distinctions are less coherent in Lavater's text than Shortland would have us believe. He defines the opposition between scientific and philosophical physiognomy in the following manner:

> Scientific physiognomy has the object of arranging, specifying and defining those stable features [as opposed to mobile features] while philosophical physiognomy is what we would today term the physiology of expression—the domain investigated by Descartes, Le Brun, James Parsons, Cureau de la Chambre and John Bulwer before Lavater, and following him by Charles Bell and Darwin. [47]

It is not at all clear in what way this distinction differs from the distinction between physiognomy and pathognomy, as both seem to be based on the opposition of fixed facial structure and mobile expression. It is equally unclear why the "physiology of expression" should be less "scientific" and more "philosophical" than the ordering of the fixed features. Shortland seems to give Lavater more credit for his intent than for his execution: if the interpretation of the stable features can indeed be said to be a goal of Lavater's work, it can by no means be said to have been accomplished. One of the items one might expect from a systematic treatise on physiognomy, as the *Essai* purports to be, is a catalogue of the facial features with some rules for interpreting them. This, however, never appears in any organized fashion in Lavater, as one might believe from reading Shortland. If we were to approach Lavater's text in good faith (that is to say, suspending all ideological cynicism), seeking advice as to the interpretation of the features according to some general guidelines, seeking in fact Shortland's arrangement, specification, and definition of the features, we would indeed be at a loss. To attribute any kind of successful systematization to Lavater is simply erroneous.

Shortland puts most of his emphasis on the final distinction he claims Lavater makes, between sensation and perception, however, and it is here that his argument is most germane to my purposes. He outlines Lavater's concept of "physiognomical perception" in the following manner:

> having defined his perceptual object as those stable parts of the body observable from the exterior, Lavater distinguishes the physiognomical sensation, which is a universal attribute of all creatures, a vague understanding that form speaks of content which underpins the majority of description, from physiognomical perception which man alone can develop and which allows him to *think* rather than feel physiognomically.[48]

Shortland goes on to say that what Lavater is introducing with his physiognomy is a way of perceiving things *"differently,"* and gives as his justification the fact that Lavater's physiognomy, unlike earlier works on the subject (Aristotle, della Porta, Le Brun), "firmly rejects" analogy (e.g., between the sexes, between the races, and between men and animals) as one of its tenets.[49] Perhaps this is true in the edition Shortland uses (the seventeenth edition, from Holcroft's English translation), but in the original French edition (largely overseen by Lavater himself) as well as other early editions, there are several examples of analogy between, for example, human faces and various animals (mostly borrowed from Le Brun) and, as we have seen above, ample evidence of analogies between men and women. Although the method of analogy is indeed not central to Lavater's project, it is plainly not one that he rejects.

The basic premise of Shortland's analysis is nonetheless that, in spite of inconsistencies, "a more unitary perspective does however arise spontaneously from the work once we view it as articulating a science of physiognomical perception."[50] His analysis would perhaps have been better served had he examined the inconsistencies more closely, and given Lavater less credit for having successfully articulated anything at all.

To be fair to Shortland, it must be said that he does indeed recognize several of the important contradictions in Lavater's work, and he is right in signaling Lavater's emphasis on observation and perception. However, he is even more right when he quotes Lavater as saying on the one hand that all men can become physiognomists, and on the other that the "good" physiognomist is an extremely rare person indeed. Shortland's reading of this seeming contradiction is that "the chosen few are the scientific practitioners, those who have . . . refined their visual perceptions . . ."[51] This fits in well with his idea that the fundamental distinction Lavater makes in defining physiognomy is that between perception and sensation. It is not, however, the case, as Lavater in fact values sensation (or at least a certain sensation) above perception, and thus undermines his very project.

At several points in the *Essai,* Lavater refers to something he calls *le sentiment physiognomonique,* which seems to be an intuitive power of interpretation unencumbered by analysis or method. The true physiognomist is predisposed to recognizing metaphysical truth from cursory observation of the body, and does not need science to feed his perceptions, as Lavater tells us: "on doit parvenir à se convaincre que le sentiment physiognomonique est la première base de la science des physionomies, qu'il est antérieur à toute expérience, à toute comparaison, et à tout raisonnement" (1: 195). Lavater uses the words *première* and *antérieur* here as terms which express not only chronology but also importance. Not only is the *sentiment* or intuition felt by the physiognomist the first step in the process of interpretation, it is also the most important—indeed, sometimes the only step. The true physiognomist, as a genius, is better served by his first subjective impression than by any methodical investigation. Lavater advises the physiognomist of this:

> Abandonnez-vous toujours aux premières impressions et même fiez-vous y davantage qu'aux observations. Vos aperçus sont-ils le résultat d'un sentiment involontaire, excité par un mouvement subit? Soyez sûr

que la source en est pure, et que vous pouvez vous passer de recourir à l'induction. (2: 374)

The source in question here is clearly not only "pure," but divine. The first impressions of a true physiognomist are God-given and natural, unlike human induction which is subject to all manner of fallacy. The physiognomist is a genius with a *God-given* talent for interpreting the signs of the human body. Indeed, he is distinguished from other men by his ability to read the "natural, original" language inscribed on the human body:

> Lui seul comprend la plus belle, la plus éloquente, la moins arbitraire, la plus invariable et la plus énergique des langues, *la langue naturelle de l'esprit et du cœur, de la sagesse et de la vertu* . . . il reconnaît la vertu, à travers tous les voiles qui la cachent. (1: 77–78)

As we saw above, Lavater defines here the language of physiognomy as "natural," essential, God-given as opposed to the arbitrary, error-prone, human one. The object of study of physiognomy (this essential, corporeal language) is therefore beyond question. The physiognomist himself, as the recipient of God-given abilities and hunches ("un sentiment involontaire"), is also beyond question. This is of course further proof that Lavater's physiognomy is primarily a theological, and not a scientific, enterprise. Not surprisingly, we find him making statements about this "genius" that directly contradict even the most elementary notions of scientific method. It is surely difficult to conceive of a science whose students are advised to trust their first impressions rather than any systematic observation; in fact, they are advised that if these impressions come from a sort of gut feeling ("sentiment involontaire"), they need not bother with induction at all.[52] Lavater's teaching here is once again in direct opposition to science, which generally defines itself as based on method, observation, and objectively verifiable data.[53]

Scientific method is, at least by traditional definitions, a refutation of the use of intuition and sentiment to formulate ideas about the world. Although absolute distinctions between science and sentiment (and between science and theology) were not yet clearly established in the eighteenth century, even contemporary readers were sensitive to the contradictions within Lavater's text itself, and his espousal of such mutually exclusive categories as intuition and observation renders his "method" somewhat less than coherent.

As we saw above, Lavater gives primacy to the involuntary, subjective, and probably inexplicable first impressions of a face that a physiognomist might have. This does not, however, prevent him from advocating the need for close and methodical observation as well:

> Le physiognomiste ne se laisse point séduire par des *apparences;* il les examine et les étudie avec soin, persuadé que chaque apparence est fondée sur une réalité. (3: 39)

> Mépriser les détails, c'est mépriser la nature. (2: 368)

The contradiction is clear: what is a physiognomist doing when he "abandons himself" to a "sentiment involontaire" about a human body, if not letting himself be seduced by appearances? When he trusts his first impressions more than his observations, as per Lavater's explicit directive, is he not necessarily obliged to "shun the details"? Once again, the reader of Lavater's "how-to" manual is given completely incompatible instructions and is left without the slightest idea "how-to." [54]

Even if the reader were to follow one set of instructions and disregard the other, he would be no more enlightened, as each contains its own particular insurmountable conceptual and practical ambiguity. For example, to consider the problem from a pragmatic point of view, a would-be physiognomist certainly has no means of creating or provoking divinely inspired intuition; genius, by definition, can not be willed. This method can not be learned and is therefore beyond his control. If he were to follow the other directive cited above ("Le physiognomiste ne se laisse point séduire . . ."), he would be obliged to try not to judge solely on the basis of appearances. If the object of study is a human body, how can he not judge on appearances? Lavater tells the physiognomist to go beyond the appearance, to study and examine carefully in order to ground his judgment not on appearance, but on "reality." The problem here is that without some attempt on Lavater's part to articulate what this reality might be, or how the physiognomist might gain access to it in order to verify his observations, one is left with the same void on the methodological level as on the conceptual. He alludes to a reality that transcends appearance at the same time as he tells us that it is through appearance that we have access to this reality, and never provides any information about how to make the leap from observation of the physical to knowledge of the metaphysical. We are left once again with the idea that one must simply be born a physiognomist.

Lavater takes this notion to a sublimely ludicrous degree when he asserts that a physiognomist must be gifted with physical as well as perceptual attributes. In a fanciful twist on the notion that "it takes one to know one," he tells us that only a beautiful man is really in a position to judge the beauty of others, and concludes that indeed it is the dearth of human physical beauty that has led to the discrediting of physiognomy:

> Sans les avantages de la figure, personne ne deviendra bon physiono-miste. Les plus beaux peintres sont aussi devenus les plus grands pein-tres . . . De même que l'homme vertueux est le mieux en état de juger de la vertu, l'homme droit de ce qui est équitable et juste, de même ceux qui ont les plus beaux visages sont les plus capables de pronon-cer sur la beauté et la noblesse des physionomies, et de découvrir en même temps ce qu'elles ont d'ignoble et de défectueux. Si la beauté était moins rare parmi les hommes, peut-être la physiognomonie serait-elle plus accreditée. (1: 117)

The passage above is perhaps the clearest illustration of the fact that physiognomy is concerned not so much with man as "sujet d'étude" as with man as "celui qui étudie." To borrow the terms of literary studies, we can say that physiognomy concerns itself with the reader and the reading process more than with the text itself. This is per-haps the only way to explain the inexplicable and assured readings of specific faces Lavater provides as examples in the *Essai*.

If the emphasis of physiognomy is on the genius of the person who practices it rather than on the accuracy of his analysis, who should illustrate this better than Lavater himself? Mirabeau describes him in the following manner in his pamphlet *Lettre à M . . . . sur M.M. Cagliostro et Lavater:*

> Ce Lavater auteur à 36 ans de 80 volumes, est peut-être un des plus sin-guliers personnages de ce siècle. On connaît en Europe les quatre tomes énormes de poésie en prose qu'il a donnés sur l'art physiognomical, et dans lesquels se montrent quelques tours de génie . . . [55]

It is true that the above passage was probably written with much irony intended, as the pamphlet is a polemic against Lavater's mysticism and Jesuitical tendencies. However, Mirabeau unknowingly provides us with a good illustration of an important phenomenon of Lavater's physiognomy, the fact that the *Essai* can be seen as deriving its inter-est from its success not as a scientific manual but as a literary text. Furthermore, it is a text which can be read as related to narrative, and raising pertinent questions about narrative. To be sure, Mirabeau's characterization of the *Essai* as "quatre tomes énormes de poésie en prose" underscores the "literariness" of the work only to discredit the scientific, "real" value to which it pretends. However, this characterization serves to remind us of the fact that the interest of Lavater's work is not necessarily diminished by the fact that it fails as a work of scientific reference.

Clearly, the focus of this study in the chapters which follow will be physiognomy as it is inscribed, thematized, and problematized in the novel. We have seen that the rhetoric of Lavater's physiognomy does not hold up to a close reading, and in fact that physiognomical thought in general can be said to be based on erroneous and contradictory premises. None of Lavater's many nineteenth-century successors in the "science" was ever able to avoid these contradictions and thereby answer Lavater's call for someone to synthesize his fragments.[56] It is nonetheless a fact of cultural history that this mode of interpretation was widely espoused by many literate Europeans of a certain period. Lavater's work is highly suggestive in relation to some of the less frequently explored elements of the novel, particularly those concerning the (fictional) physical selves of characters and the significant ways in which these fictional physiques are verbally depicted. Despite its systemic failure, therefore, it is of potential importance to a student of eighteenth and nineteenth-century cultural and literary studies. What is important to me is how this mode of thought is reflected in both *récit* and *histoire* of certain French novels of that same period.

Perhaps the best place to begin a discussion of physiognomy in relation to narrative would be with the narrative of physiognomy itself. We have already seen physiognomy as a linguistic construct, subject to the inherent vagaries and inadequacies of language. We have also seen that physiognomy, specifically that of Lavater, is useless as any kind of scientific reference or manual of interpretation of the human form. What is a physiognomical reading then, and in what terms can it be defined? When Lavater examines an engraving, and proceeds to give a description of the physical features it displays, while also giving an account of the character displayed by these features, he is not relating the results of an accurate, verifiable scientific investigation based on a sound conceptual base. Nonetheless, he is recounting something. The act of interpretation effected by Lavater, as is true of any act of interpretation, is the creation of a narrative. As any reader of literature knows, the interest of a narrative is not necessarily dependent on its referential accuracy, and this can be true of texts which pretend to this accuracy (like the *Essai*) as well as of explicitly fictional texts.

In reading the narratives of faces that Lavater gives in his work, one is frequently reminded of the popular pastime that consists of creating imaginary biographies of passersby, based solely on the few clues one can garner from their appearance.[57] The narratives Lavater creates are as devoid of veracity as those created in this game. They are, however, unlike those of the game, not without purpose: they are attempts (albeit failed) to explain and to order human being.

In his relevant discussion of "plot," *Reading for the Plot*, Peter Brooks makes the following statement about plot and narrative: "Plot, let us say in preliminary definition, is the logic and dynamic of narrative, and narrative itself a form of understanding and explanation."[58] If, according to this definition, narrative can be characterized as that which imparts intelligibility,[59] the relation between narrative and physiognomy becomes quite clear. As an effort to deny the arbitrary, physiognomy is above all an attempt to order, to explain, to narrate. In fact, Brooks outlines the relation between the social and natural sciences (among whose number physiognomy is eager to be included) and narrative-as-intelligibility:

> Not only history, but historiography, the philosophy of history, philology, mythography, diachronic linguistics, anthropology, archaeology, and evolutionary biology all establish their claims as fields of inquiry, and all respond to the need for an explanatory narrative that seeks its authority in a return to origins and the tracing of a coherent story forward from origin to present.[60]

It is particularly pertinent to note that Brooks writes the above passage after having reminded us of Voltaire's idea (which "the Romantics confirmed") that "history replaces theology as the key discourse and central imagination . . ."[61] As we saw above in an attempt to characterize Lavater's physiognomy, it is at bottom an unsuccessful conflation of theology and science. Lavaterian physiognomy attempts to exploit the narrative order offered by both discourses. Because it lacks the seemingly objective components of the disciplines listed by Brooks as substitutes for theology, it is left with little credibility as a referential discourse; conversely, however, as the intersection of two "masternarratives" (theology and science), it remains a suggestive space for inquiry into narrative and especially narrative-as-ordering.[62]

As we saw above, physiognomy has no real object: Lavater provides neither method, nor system, nor explanation of how one moves from observations on the physical to assertions on the moral. The work of physiognomy is therefore not the creation of a narrative based on observation and method, but rather the creation of a narrative based on *génie* or imagination. Lavater describes on one level the physical appearance of the face in question while creating on another level an account of the metaphysical quality shown by each feature, an explanation of each feature. Lavater's physiognomical readings are both description and narration, and therefore undo the traditional notion in literary studies that the two categories are mutually exclusive.[63]

The readings are at once fictional and explanatory: he seems to ignore the need for verifiable data and method, and forges ahead in the belief that the narratives he creates are nonetheless valid means of rendering human bodies intelligible. Although he was surely not conscious of it, Lavater's work seems to function on the principle that there is something in the nature of narrative itself that lends a coherence that is not accountable to criteria of referential veracity. This view of the *Essai* provides a clear link to a discussion of fictional discourse, since physiognomy as a topos in the novel often serves to propel and order the narrative (particularly when the novel is the story of a fictional life).

In spite of claims to science (particularly by Balzac and Zola), the "realist" or "naturalist" narrative owes its coherence not to reality or nature, but to narrative itself. Like Lavater's text, these texts are persuasive in spite of a lack of any real claim to either scientific or mimetic accuracy; they are persuasive because of the satisfaction they provide for the basic human need to order and explain life.[64]

Brooks tells us that "from sometime in the mid-eighteenth century through to the mid-twentieth century, Western societies appear to have felt an extraordinary need or desire for plots, whether in fiction, history, philosophy, or any of the social sciences, which in fact largely came from the Enlightenment and Romanticism."[65] This historical view of the desire for plot, and thus for an ordered and ordering narrative, obviously fits in well with the agenda of this study from both historical and conceptual points of view. The idea of a need for narratives of intelligibility is equally clearly illustrated by Lavater's attempted theory and the use of physiognomical characterization in novels to order and explain the events in the lives of the characters.[66] It has indeed been suggested that, after Lavater, it was fictional, narrative literature (such as the *Comédie humaine*) which perpetuated the physiognomical tradition.[67] Indeed, as I shall discuss in the conclusion, the literary works here represent an inscription of the conceptual and rhetorical gestures of the tradition of physiognomical discourse per se, as exemplified by Lavater.

If, by definition as narrative, these novels attempt to explain human existence and are thereby analogous to Lavater's project, they also share in its rhetorical failure. Each of the novels I discuss in this study, like Lavater's *Essai*, fails to provide its reader with sufficient information for cracking the code of physiognomical knowledge its plot suggests. The reader of these novels, like the reader of Lavater's treatise, is excluded from the knowledge necessary to fully comprehend and interpret the text he/she is reading.

In the following chapters, then, which treat works of narrative fiction which postdate Lavater, an understanding of the various versions of physiognomical thought as a semiotic and narrative system will continue to serve as an *entrée en matière* to analyses of physiognomy as theme, narrative device, and figure of interpretation, as well as to questions concerning physiognomical definitions of gender.

Although these questions are germane to all of the discussions of literary works in this study, nowhere are they as explicitly linked to Lavater than in the following chapter on Balzac. As we shall see, Balzac's version of physiognomy, while apparently (if sometimes, perhaps, ironically) remaining faithful to Lavaterian principles, represents a conscious effort on the part of its author to expand and modify the Lavaterian object of interpretation by the inclusion of such social signifiers as *toilette, démarche,* and *vestiognomie.*

Chapter Four

Balzac and *l'homme hiéroglyphié*

Pour mériter les éloges que doit ambitionner tout artiste, ne devais-je pas étudier les raisons ou la raison de ces effets sociaux, surprendre le sens caché dans cet immense assemblage de figures, de passions et d'événements?

—Balzac

The topic of Balzac and physiognomy is hardly a novel one: indeed, it has been noted and commented on by many critics in the past century. The fact that Balzac is often quite explicit about his debt to Lavater has paved the way for a plethora of thematic studies on the influence of the "science" on the Balzacian novel. Unfortunately, with almost no exception, these studies are statistical, facile, and simply dull. They generally attempt to prove what has already been proven (that Balzac read Lavater, and that his literary portraiture bears the mark of this reading) through long-winded catalogues of the descriptions of various physical features throughout the *Comédie humaine*. Perhaps it is the sheer volume of the *Comédie* that leads some of these critics to believe that the real job of the critic of Balzac is to scour the immense *œuvre* and compile a detailed list of the myriad appearances of whichever theme happens to be his/her particular concern. Although this can be in itself an impressive feat, such a list nevertheless fails to perform its own "physiognomical" reading of the texts it treats, and we are left with an accounting of only the most superficial textual phenomena.

There are, however, important questions raised by these themes if we attempt to interpret their functions and implications in the texts in which they are *mis en scène*, rather than merely acknowledge their presence. By limiting the analysis in this chapter to a few important texts by Balzac and reading them both separately and through each other, I hope to escape the urge to catalogue and overgeneralize by analyzing specific cases of the function of physiognomy as an ideological and aesthetic phenomenon in Balzac.

In spite of a seeming contradiction with the goals stated above, it would be both pertinent and helpful to begin with a brief overview of the information presented by earlier critics on my topic, so as to provide a general context for my analysis. Most critics establish the fact that Balzac owned a copy of Lavater's essay on physiognomy by quoting a letter he wrote to his sister (20 August 1822): "J'ai acheté un superbe Lavater qu'on me relie . . ."[1] In addition to this bit of biographical information, there are numerous explicit references to Lavater in Balzac's work. Indeed, Pierre Abraham has estimated that Balzac cites Lavater or Gall more than one hundred times in the *Comédie humaine*.[2] An oft-cited passage in which Balzac pledges allegiance to Lavaterian physiognomy is in the *Physiologie du mariage* (Méditation 15):

> La physionomie a créé une véritable science. Elle a pris place enfin parmi les connaissances humaines . . . Les habitudes du corps, l'écriture, le son de la voix, les manières, ont plus d'une fois éclairé la femme qui aime, le diplomate qui trompe, l'administrateur habile ou le souverain obligés de démêler d'un coup d'oeil l'amour, la trahison ou le mérite inconnus. L'homme dont l'âme agit avec force est comme un pauvre ver luisant qui, à son insu, laisse échapper la lumière par tous ses pores . . .[3]

There can be little doubt about Balzac's having read Lavater and believed in the basic precept of his theory, that man's "interior" self can be deciphered from clues given by his "exterior" self. It is probably accurate, in fact, to say that of all nineteenth-century European novelists, Balzac is the one whose work is most clearly influenced by the then widely known theory of physiognomy.[4]

As Pierre Abraham has noted, Balzac frequently cites not only Lavater but also Franz Joseph Gall, creator of the well-known subset of physiognomy known as phrenology.[5] Phrenology narrows the field of physiognomical practice by taking as its sole object of interpretation a very specific set of corporeal signifiers, those of the conformation of the skull. The shape and proportion of the various areas of the cranium are read as indices to moral and intellectual propensities. Although phrenological practice is thus somewhat different from that of Lavaterian physiognomy, their underlying principles are identical: the analysis of character based on corporeal signifiers. Indeed, Gall's most fundamental and original gesture, the choice of the cranium as the object of physiognomical interpretation, reflects above all a specifically Lavaterian emphasis on the importance of immutable, morphological signs as opposed to voluntary, mobile features.

Because I am much more concerned in this study with the theoretical implications and literary inscriptions of physiognomical thought than with the practical applications of its many variants, I will not discuss in any detail phrenology as such. With respect to Balzac, it is important to note that, for all of his frequent citations of Gall, there are, not surprisingly, no instances of specifically phrenological themes in his oeuvre. More often than not, Balzac mentions Lavater and Gall or physiognomy and phrenology in the same sentence, without specifying that the relation between the two is defined by the subordination of phrenology to physiognomy.

If Balzac's specific debt to Gall is insufficient to warrant thorough analysis, the same cannot necessarily be said of his debt to Lavater. After having established the basic facts about Balzac and physiognomy outlined above, critics are faced with the more difficult task of identifying the manner in which Lavater's influence manifests itself in Balzac's work. Henri Gauthier outlines some specific analogies between Lavaterian physiognomy and Balzacian character description, including the division of both the body and the face into three zones, each with its own significance; the structure of analysis which takes as its objects the stature, head, and face, respectively; and the notion of the human body as a unity, the essence of which is discernible in any one of its component parts.[6]

Fernand Baldensperger cites a number of passages in which Balzac would seem to be following specifically Lavaterian directives, both in according more importance to certain physical features than to others, and in the specific interpretations of these features. Indeed, the correspondence between the passages Baldensperger cites from Lavater and those from Balzac is striking: for example, Lavater says that "De grosses lèvres bien prononcées répugnent à la méchanceté," while Balzac reports that Mlle. Cormon (the title character of *La Vieille Fille*) has "de grosses lèvres rouges, l'indice d'une grande bonté." As for the chin, Lavater states that "un menton mou, charnu et à double étage est, la plupart du temps, la marque et l'effet de la sensualité"; Balzac, in describing Véronique in *Le Curé de village,* informs us that "l᷾ menton et le bas du visage étaient un peu gras, et cette forme épaisse est, suivant les lois impitoyables de la physiognomonie, l'indice d'une violence quasi-morbide dans la passion."[7] Baldensperger ends his exposition of these analogies by demonstrating that the ear is considered to be of little importance by either author.

For his part, Gilbert Malcolm Fess cites Balzac's debt to Lavater in the areas of human resemblance to animals, coloration, eye color, hair, general body shape, hands, noses, and teeth, among others.[8] Fess goes on to list areas of physiognomical "theory" that are of Balzac's own creation, not directly patterned after Lavater, including theories specifically concerning hunchbacks, social class, geography, eyebrows, facial hair on women, necks, hips, and feet.[9] There are also areas which Balzac explores in greater depth than Lavater, objects of analysis to which Lavater merely refers but about which Balzac obsesses: gait (this is the basis of his *Théorie de la démarche*), gesture, voice, comparisons between men and animals, clothes, and the effect of *milieu* in general.[10] Fess summarizes his view of Lavater's influence on Balzac by saying that it was not so much the specific details of Lavaterian physiognomy as its basic precepts that influenced Balzac:

> The influence of Lavater on Balzac was not, however, limited to concrete information. In the first place, his ill-arranged and obscure statements sometimes furnished the latter with a starting point from which to reach conclusions of his own by strictly logical processes. In the second place, he inspired the author of the *Comédie humaine* to observe the humanity about him and study the relationship of character to physical form . . . This, certainly, was his most significant contribution.[11]

Pierre Abraham, in *Créatures chez Balzac*, says that Balzac's method of physical description of characters is one of "correspondance psycho-physiologique" and that we should logically expect it to follow some sort of "tableau de correspondance" between physical features and character traits furnished by Lavater (and/or Gall). Instead, Abraham tells us, largely because of the lack of cohesion in Lavater's own work, Balzac merely takes his basic premise from Lavaterian physiognomy and creates his own "tableau de correspondance."[12]

Abraham goes on to outline his version of Balzac's "tableau" in some detail. He points out that certain hair colors, eye colors, and combinations of the two are greatly overrepresented in Balzac when compared with anthropometric studies of the same features in the French population. Using this as his proof that Balzac is indeed more interested in creating his own system of correspondence between hair and eye color and character than in strict statistical realism, Abraham gives us approximately the following bases of Balzac's "tableau": blue eyes signify activity, brown eyes passivity, blond hair independence and strength, and dark hair passion, love, and sexuality. Less usual is red hair, which signifies bestiality. Abraham posits the general idea that for Balzac, quantity of pigmentation is directly proportional to quantity of passion.

If the above-mentioned critics agree that Balzac was at least to some extent influenced by Lavater (and, by extension, Gall) in his physiognomical thought and that this influence makes itself clear at various points in his work, they also survey the aspects of Balzac's thought which are physiognomical (in that they take as their premise a legible correspondence between physical and metaphysical man) without having been influenced by Lavater. I should specify that by this I am not referring to the differences on the level of specific corporeal detail cited above, but rather to aspects of Balzac's physiognomical thought which draw on sources unrelated to Lavater and physiognomy per se.

Perhaps the most fundamental (and no doubt the strangest) of Balzac's theories about the correspondences between physical and metaphysical man is that of the existence of a "vital fluid," a mysterious liquid which serves as a conduit between body and soul. The origin of this theory is clearly Mesmer's theory of "animal magnetism." In *Ursule Mirouët*, Balzac explains that magnetism is the very base of both physiognomy and its corollary, phrenology:

> La phrénologie et la physiognomonie, la science de Gall et de Lavater, qui sont jumelles, dont l'une est à l'autre ce que la cause est à l'effet, démontrent aux yeux de plus d'un physiologiste les traces du fluide insaisissable, base des phénomènes de la volonté humaine, et d'où résultent les passions, les habitudes, les formes du visage et celles du crâne.[13]

Henri Gauthier explains the close relation between the occult sciences of magnetism and physiognomy in the following passage on *Ursule Mirouët*:

> Balzac fait dans ce roman un exposé sur le Mesmérisme, ou plutôt sur le magnétisme dont Mesmer n'a pas su établir la véritable science. Le magnétisme se rattacherait à la plus haute antiquité: il fut cultivé "par l'Egypte et par la Chaldée, par la Grèce et par l'Inde"; il fut "la science favorite de Jésus et l'une des puissances divine remises aux apôtres." Par cette évocation qui complète les vues historiques sur la religion exposée dans la préface du *Livre mystique* et qui seront reprises en 1842 dans l'édition Charpentier de *Louis Lambert* . . . les sciences occultes fondées sur le magnétisme, se rapprochent grâce à leur source commune. Ainsi la magie et le miracle s'expliquent naturellement par "la science des fluides impondérables, seul nom qui convienne au magnétisme si étroitement lié par la nature de ses phénomènes à la lumière et à l'électricité," tout comme la science de Gall et Lavater . . .[14]

Although neither Balzac nor any other proponent of the theory of animal magnetism ever defines the chemical makeup or anatomical function of the mysterious fluid, it is clear that the theory is an attempt to literalize a relation between physiology and spirituality, between the body and the spirit.[15]

The most significant consequence of the "vital fluid" theory for Balzac was that it led him to believe that the fluid represented an *essence* of life, an essential unit which linked human and animal physiologies. With this fundamental link established, Balzac could go on to draw further, more detailed, analogies between the two "kingdoms." Because the essence of life had been defined as being a single substance, Balzac believed that the real work of the "scientist" was to identify the similarities and differences among creatures, among the various external forms the "essential substance" could adopt. The next step in Balzac's method was to establish classifications based on these comparisons. One could then theorize about these classes in a general way, and individuals belonging to a class (for example, specific characters in a novel) could be characterized by reference to the classification. The model Balzac uses in this process was a very topical one in the first half of the nineteenth century, that of zoology.

No critic defines the relation between Balzac's agenda in the *Comédie humaine* and the theory of zoology more clearly than does Balzac himself, in his famous Avant-propos to the *Comédie* (1842):

L'idée première de *la Comédie humaine* fut d'abord chez moi comme un rêve . . . Cette idée vint d'une comparaison entre l'Humanité et l'Animalité . . .

Il n'y a qu'un animal. Le créateur ne s'est servi que d'un seul et même patron pour tous les êtres organisés. L'animal est un principe qui prend sa forme extérieure, ou, pour parler plus exactement, les différences de sa forme, dans les milieux où il est appelé à se développer. Les Espèces Zoologiques résultent de ces différences. La proclamation et le soutien de ce système, en harmonie d'ailleurs avec les idées que nous nous faisons de la puissance divine, sera l'éternel honneur de Geoffroi Saint-Hilaire, le vainqueur de Cuvier sur ce point de la haute science . . . [16]

It should be clear from the passage above that Balzac's agenda in writing the *Comédie* was a sociological one patterned after the model furnished by contemporary zoology. It was furthermore the indirect descendant of the animal analogies which were an integral feature of physiognomy from pseudo-Aristotle through Porta and Le Brun to Lavater himself. Balzac's theoretical comparison of humans and animals transcends the formalistic and occult facial resemblances found in those earlier works, however. His goal was rather to establish a series of social classifications of humans analogous to those provided for animals by zoology. Once again, his Avant-propos provides us with a clear definition of both his intent and his debt to natural history:

> Pénétré de ce système bien avant les débats auxquels il a donné lieu, je vis que, sous ce rapport, la Société ressemblait à la Nature. La Société ne fait-elle pas de l'homme, suivant les milieux où son action se déploie, autant d'hommes différents qu'il y a de variétés en zoologie? . . .
>
> Il a donc existé, il existera donc de tous temps des Espèces Sociales comme il y a des Espèces Zoologiques. Si Buffon a fait un magnifique ouvrage en essayant de représenter dans un livre l'ensemble de la zoologie, n'y avait-il pas une œuvre de ce genre à faire pour la Société? (9–10)

Balzac's debt to the natural history and zoology of his time is a fact well established and commented in traditional criticism on Balzac. Baldensperger gives the following, slightly cynical, interpretation of the role of these sciences in Balzac's system of thought:

> Le Muséum d'histoire naturelle—où un laïque pouvait risquer des incursions que rebute peut-être la technique de la méthode, mais que récompensent des vues grandioses sur le Monde—a été un des pôles de la vie cérébrale de Balzac à l'âge où prenait pour lui un intérêt singulier tout ce qui jetait quelque lumière sur la condition réelle de l'humanité. Les idées des grands naturalistes français offraient une sorte d'armature et de garantie à des notions plus audacieuses qui sont, pour cet autodidacte, les "clefs" de l'observation, de l'interprétation, appliquées à l'humanité.[17]

The above passage by Baldensperger gives us an insight not only into the biographical significance of the zoological model, but more important into its fundamental significance to the conceptual framework of the *Comédie*. Balzac's debt to zoology, like his debt to physiognomy, lies less with any specific formal influence than with more general theoretical concepts which he uses as means to his own (literary) ends. The notion of classification of different species, institutionalized by the nascent discipline of zoology, allowed Balzac to create an analogous system for humans, whereby the comparison and contrast of different groups of people could afford great opportunities to systematize the acts of observation and interpretation of human beings.

Although it is certainly useful to recognize both Balzac's pretension to social science and his debt to natural science, it is perhaps even more revealing to see where the two disciplines diverge from each other. Balzac agrees with Geoffroy Saint-Hilaire that the varying physical forms (i.e., species) that life can assume are due to *milieu*, to the physical environment in which a species exists. It is here that a direct analogy between humans and animals begins to falter, however. For animals, *milieu* is dictated by irrefutable facts of nature and is thus a geographical, climatic, and ecological phenomenon. Human society, on the other hand, organizes itself according to historical and sociological directives which create various physical environments for various sociological groups. These physical environments in turn create distinct (to the trained eye) physical signifiers which reflect affiliation with a social, economic, or even moral classification.

Although the basic animal/human analogy holds true, Balzac's attempt to typify and classify human beings proves to be considerably more complex and nuanced than most of the nineteenth-century zoological classifications from which he drew his inspiration. Peter Demetz defines the transformation of zoological principles into sociological ones, and suggests the consequent implications for literary representations of these principles:

> The very concept of the type rests on the zoological assumption that the original *dessein* is "individualized" by a continuing pressure in which the forces of time and *les milieux* combine. One immediately suspects that *les milieux*, which to the scientific mind connote the two spheres of water and air (at least in French zoology of the 1820s and 1830s), assume a far richer and variegated meaning when Balzac changes zoological into sociological relevance and speaks of man "dans les milieux où son action se déploie"; the zoological dual turns into a sociological plural of infinite potentialities . . .[18]

This "plural of . . . potentialities" is enacted through the vast number of characters and plots in the *Comédie* itself. Physiognomy and related fields of study, including the theory of animal magnetism and a sociology patterned after zoology, provided Balzac with tools of analysis for his literary dissection of man and society as well as a means of validating these attempts to theorize about human beings and their bodies. In both fictional and theoretical discourse (and, in some cases, in writings which blur the distinction between the two), Balzac uses these disciplines as the conceptual underpinnings on which his own authority as narrator and interpreter of human life rests. His agenda is not merely to represent his society faithfully in a one-dimensional, "realistic" way; it is rather to use the theories represented by various disciplines as a means of explaining human lives. In the broadest sense of the term "physiognomical," Balzac's very purpose in constructing the vast universe of the *Comédie* can be said to be based on the most fundamental physiognomical principle, one which recognizes superficial, physical phenomena as important in their capacity as signifiers. If the novelist paints a portrait of his society in all its physical detail, it is because each detail is capable of revealing the individual moral composition, essential metaphysical nature, or sociological identity of a character. Balzac sums it all up in his Avant-propos:

> S'en tenant à cette reproduction rigoureuse, un écrivain pouvait devenir un peintre plus ou moins fidèle, plus ou moins heureux, patient ou courageux des types humains, le conteur des drames de la vie intime, l'archéologue du mobilier social, le nomenclateur des professions, l'enregistreur du bien et du mal; mais, pour mériter les éloges que doit ambitionner tout artiste, ne devais-je pas étudier les raisons ou la raison de ces effets sociaux, surprendre le sens caché dans cet immense assemblage de figures, de passions et d'événements? . . .
> La loi de l'écrivain, ce qui le fait tel, ce qui, je ne crains pas de le dire, le rend égal et peut-être supérieur à l'homme d'état, est une décision quelconque sur les choses humaines, un dévouement absolu à des principes. (15–16)

Physiologies, *Le Traité de la vie élégante*, *La Théorie de la démarche*

While Balzac's unique contribution to literary history is no doubt his mise-en-scène through fictional narrative of theoretical principles such as those stated above, he also wrote texts which are some approximation of theory itself. These texts are generally short, tongue-in-cheek attempts to interpret or classify human beings by exterior phenomena such as dress, posture, and gait.

In the 1830s, Balzac wrote prolifically for various Parisian periodicals. During this period, he began to develop what would become a veritable journalistic "fad," the genre of the *Physiologie*. The *Physiologie* is usually a short, semicomic text which attempts to describe a sociological type, in much the same way that a zoologist might describe a certain species of animal, using overt physical characteristics and patterns of social behavior to serve as the defining principles of a classification. The better-known examples of the genre include the *Physiologie de l'employé* and the *Physiologie du rentier*, each of which includes numerous subdivisions such as "Le Célibataire," "Le Taciturne," "Le Philanthrope," and "Le Campagnard."[19]

The comic element of these writings derives in large part from the application of the tools and language of the analysis and classification of animals to the analysis and classification of humans. The seeming absurdity of classifying humans as if they were zoological specimens is compounded by the fact that zoological means of classification *can* be imposed on humans; some seemingly trivial human characteristics (details of dress, for example) can indeed be shown to be potentially significant when presented after the fashion of zoology. If our knowledge of animals and our ability to classify them come from such indices as body type, mating habits, and geographical location, why should the same indices not allow for classifications of humans? These sociological writings intentionally exaggerate the sense of absurdity we experience before this kind of analysis, and thus might appear to be simple parody.

However, a further, more ambiguous comic effect is produced by the tension between the reader's primary reaction (that such writings are necessarily absurd) and his secondary reaction (that, after all, what he is reading does in some way describe a certain sociological reality). The final effect on the reader of these texts is probably the laughter of uncertainty: the *Physiologies* are clearly intended to be parodies, but they are usually also conveying information which can not be dismissed as patently false.[20] One laughs because they manage to be simultaneously absurd and accurate. The *Physiologies* reflect the uncertainty of a discipline struggling to define and establish itself, to ascertain (in the case of Balzac, through the use of parody) whether a discourse which studies human beings in much the same systematized manner as zoology studies animals could overcome the potential for absurdity of the inevitable comparison between the two. It is by no means a mere coincidence that Auguste Comte coined the term *sociologie* in 1838, at the dawn of the popularity of the *Physiologies*.[21]

The basic theory which supports the *Physiologies* is shared by both physiognomy and zoology: that a creature can be analyzed and/or classified according to external qualities. That is to say, man is legible, knowable from visual indices, and the "scientist" or, to use Balzac's term, "observateur" need not interact with his subject or even observe him over a long period of time: this is a theory which equates appearance with reality, surface with depth. While Balzac puts this way of interpreting the world into practice in the various *Physiologies*, it is of even greater interest to our study to look at the texts in which Balzac claims to expose the theories on which his practice is based. Although these texts pose as serious works of theoretical, even scientific, discourse, they, like the *Physiologies*, satirize both other works of theory and their own stance as theory while communicating ideology and methodology which Balzac took with some (indefinable) measure of seriousness.

The *Traité de la vie élégante* is, as one would expect from its title,

Balzac's treatise on the importance of clothes in the observation and interpretation of one's fellow man. Balzac expounds on the fundamental importance of the observation of exterior man:

> Un traité de la vie élégante, étant la réunion des principes incommutables qui doivent diriger la manifestation de notre pensée par la vie extérieure, est en quelque sorte la *métaphysique* des choses.[22]

A bit further on, he makes his understanding of the relation between man as he appears and the reality of his being even more explicit:

> La vie extérieure est une sorte de système organisé, qui représente un homme aussi exactement que les couleurs du colimaçon se reproduisent sur sa coquille. Aussi, dans la vie élégante, tout s'enchaîne et se commande. Quand M. Cuvier aperçoit l'os frontal, maxillaire ou crural de quelque bête, n'en induit-il pas toute une créature, fût-elle antédiluvienne, et n'en reconstruit-il pas aussitôt un individu classé, soit parmi les sauriens ou les marsupiaux, soit parmi les carnivores ou les herbivores? . . . Jamais cet homme ne s'est trompé: son génie lui a révélé les lois unitaires de la vie animale.
>
> De même, dans la vie élégante, une seule chaise doit déterminer toute une série de meubles, comme l'éperon détermine un cheval. Telle toilette annonce telle sphère de noblesse et de bon goût . . . Jamais les Georges Cuvier de l'élégance ne s'exposent à porter des jugements erronés . . . Cet ensemble rigoureusement exigé par l'unité rend solidaires tous les accessoires de l'existence; car un homme de goût juge, comme un artiste, sur un rien. Plus un ensemble est parfait, plus un barbarisme y est sensible. (192)

In the passage above, with the analogy to Cuvier, Balzac reprises the notion, already exposed in the Avant-propos, of the zoological model which serves as the basis of his sociology. What interests us here is not so much the fact of interpretation and classification of man as the specifics of the semiotic system outlined in the *Traité*. In the passage above as well as in the Avant-propos, Balzac clearly outlines a conceptual justification of his mode of analysis without defining the means of its practice. By contrast, the rest of the *Traité de la vie élégante* serves as an illustration of a specific set of indices, those furnished by clothes. Balzac tells us in characteristically hyperbolic style that clothing reveals "all" about its wearer:

> *La toilette est l'expression de la société.* Cette maxime résume toutes nos doctrines et les contient si virtuellement, que rien ne peut plus être dit qui ne soit un développement plus ou moins heureux de cet aphorisme . . .

> Pourquoi la toilette serait-elle donc toujours le plus éloquent des styles, si elle n'était pas réellement tout l'homme, l'homme avec ses opinions politiques, l'homme avec le texte de son existence, l'homme hiéroglyphié? Aujourd'hui même encore, la *vestiognomie* est devenue presque une branche de l'art créé par Gall et Lavater. (209–11)

Here Balzac defines his agenda as that of furthering a branch of physiognomy, of taking the Lavaterian principles of a semiotics of the body and exploring them in relation to a specific set of signifiers unexplored by Lavater himself. He characterizes clothing as "éloquent," and in so doing reaffirms Lavater's notion that exterior signs constitute a language which communicates more effectively than that of words. Sartorial signifiers constitute a "texte" which must be read and interpreted; indeed, man's clothing inscribes upon his body hieroglyphs, signifiers which may be meaningless form to some, but are in fact a reliable code of meaning to those who know how to read them.[23]

If Balzac defines a specific set of signifiers, he also defines their correspondent signifieds. Clothes do not reveal arbitrary information about their wearer: they reveal his or her social status. This would seem logical, as the type of clothing one wears is usually dependant to some degree on the amount of money at one's disposal, and, in nineteenth-century France as elsewhere, social status was more often than not linked to economic status. However, Balzac's sartorial hieroglyphs are much more nuanced than that facile criterion; clothes reveal taste, education, background, and other subtleties of affiliation with social and professional classifications:

> Quoique, maintenant, nous soyons à peu près tous habillés de la même manière, il est facile à l'observateur de retrouver dans une foule, au sein d'une assemblée, au théâtre, à la promenade, l'homme du Marais, du faubourg Saint-Germain, du pays Latin, de la Chaussée d'Antin; le prolétaire, le propriétaire, le consommateur et le producteur, l'avocat et le militaire, l'homme qui parle et l'homme qui agit.
>
> Les intendants de nos armées ne reconnaissent pas les uniformes de nos régiments avec plus de promptitude que le physiologiste ne distingue les livrées imposées à l'homme par le luxe, le travail ou la misère. (211)

It is important to note here Balzac's choice of the term "physiologiste" to refer to the observer/interpreter of human life. The very structure of this word is one which can be interpreted as accurately reflecting the discipline outlined by Balzac in such texts as the *Traité:* as Balzac himself states, his is a discipline which takes its primary tenets from physiognomy but which combines these foundations with the practices of what can only be defined (at least in historical retrospect) as sociology. Hence the word *physiologie*, which recalls both disciplines ("physio-" echoing *physiognomonie* and "-ologie" echoing *sociologie*), is an overdetermined choice of name for a very particular field of study devised by Balzac to put the tools of physiognomy into practice to serve the goals of sociology.[24] Whereas Lavaterian physiognomy, with a theological foundation, analyzes for the most part God-given corporeal traits which reflect God-given moral or characterological traits, Balzac's *physiologie* often emphasizes the analysis of man-made physical attributes (such as clothes or, as we saw above in reference to the *Physiologies*, milieu) to reflect man-made typologies (such as social class or professional identity). The *Traité* differs from the various *Physiologies* in that it presents itself as theory, the exposition of one of the principal methods of arriving at "physiological" analysis, rather than as an example of such analysis itself.

Although one can not say that Balzac is consistent in his break from Lavaterian physiognomy (he frequently uses fixed physical features and essentialist character analysis as well), we can say that there is a very important distinction to be made between the objects of analysis of physiognomy and those of Balzac's *physiologie*. Indeed, Balzac is quite clear about both the signifier and the signified of his semiotic pursuit in the *Traité* when he says (and emphasizes) that "la toilette est l'expression de la société."

We said above that the *Traité* posits itself as a theoretical work, one which should logically enlighten its reader about the workings of *physiologie*. In fact, it is as farcical as any of the *Physiologies* themselves. The self-parody begins on the very first page when Balzac states his desire to write a treatise that will be understood, but undercuts his lofty-sounding agenda by identifying his enemies by the type of boots they wear ("Nous tâcherons, en professant les doctrines les plus secrètes de la vie élégante, d'être compris même de nos antagonistes, les gens en bottes à revers"). This very odd and amusing opening sets the tone for a very odd and amusing piece of writing, as it reflects the ambiguity and inconsistency inherent in all of Balzac's "physiological" writings: we can read the "gens en bottes à revers" remark as being an absurd parody of the very subject of the treatise (as it would certainly appear at first glance), or we can accept Balzac's theories about the all-revealing nature of clothing and find the wearing of a certain style of boot to be perhaps justifiable grounds for enmity. Balzac leaves us with the tension created by our inability to give a definitive reading to this sentence and to his text as a whole.

The *Traité* is composed of three parts, each of which in turn comprises numerous aphorisms interspersed (in a seemingly random fashion) with sometimes lengthy anecdotes and generalities, such as those I quoted above, all more or less on the topic of *toilette*. There are virtually no satisfactory conceptual transitions from one subdivision of the text to another. Although one can certainly piece together a sort of theory of the semiotics of clothing from this text, as we did above, it is of primary importance to take into account the sort of "fort-da" rhetorical style employed by Balzac to expose this theory. The reader is always kept off balance, never quite sure whether the entire text is intended as parody of theory or as theory itself. It is as if the theoretician is himself not quite certain about the validity of his theories, so he renders them just absurd enough to allow himself to deny responsibility for them. The reader is left with the impossible task of trying to construct his own theory of the theory from the intentionally ambiguous clues provided by the text.

The *Traité de la vie élégante* is not the only text of absurdist theory written by Balzac. There are also shorter texts which are best categorized with the *Physiologies* because of both their specificity and their brevity such as the *Nouvelle Théorie du déjeuner*. There is, however, one other text which is analogous to the *Traité* because of its stated intention to carry on Lavater's work in a direction only briefly mentioned by him, and thus to open up an entire new direction of "physiological" exploration. The *Théorie de la démarche* is indeed the most often cited example of Balzac's tongue-in-cheek theoretical writings; many critics are familiar with it, but few seem to know what to say about it.

The *Théorie* follows many of the patterns we saw above in the *Traité* and sets out the same agenda for itself: exposing one of the ways in which the body can be rendered an intelligible nexus of signification. Indeed, as Peter Brooks tells us, it is in the *Théorie* that Balzac most explicitly sets forth the quintessentially Lavaterian idea (also expressed in the *Traité*) that "nothing is meaningless, that the world cannot *not* mean . . ."[25] The text itself is made up of a vertiginous interweaving of axioms, anecdotes, and general musings from which a semblance of a theory of the semiotics of bodily movement can be gleaned. The text is rife with the same hyperbole, self-parody, and ambiguity found in the *Traité*. Balzac bases his theory on the same Lavaterian principles of synecdoche (any part reveals the essential nature of the whole) and physico-characterological correspondence as well:

> Lavater a bien dit, avant moi, que, tout étant homogène dans l'homme, sa démarche devait être au moins aussi éloquente que l'est sa physionomie; la démarche est la physionomie du corps. Mais c'etait une déduction

naturelle de sa première proposition. *Tout en nous correspond à une cause interne.* Emporté par la vaste cours d'une science qui érige en art distinct les observations relatives à chacune des manifestations particulières de la pensée humaine, il lui était impossible de développer la *Théorie de la démarche,* qui occupe peu de place dans son magnifique et très prolixe ouvrage. Aussi les problèmes à résoudre en cette matière restent tout entiers à examiner, ainsi que les liens qui unissent cette partie de la vitalité à l'ensemble de notre vie individuelle, sociale et nationale.[26]

With his intention to continue Lavater's work so clearly defined, it should come as no surprise that the theory put forth in the *Théorie,* in contrast to the *Traité de la vie élégante,* emphasizes the physiognomical as opposed to the sociological. Balzac concerns himself with bodily movement as it indicates character rather than social class.

As far as the specific content of Balzac's theory of movement is concerned, it is fairly predictable and can best be summarized by a series of axioms, both explicit and implicit, interspersed throughout the text. Judith Wechsler provides us with a concise and accurate summary of these axioms:

> The walk announces the man; gesture is thought in action, by which one can decipher vice, remorse, sickness; the look, the voice and walk are equal means by which you can know the entire man; all our body participates in movement, but no part should predominate; when the body is in movement, the face is immobile; economy of movement is the means to render a noble and gracious walk; all jerky movement betrays a vice or a bad education; grace favours rounded forms; rest is the silence of the body; work has its effect on our bearing, scholars, for instance, incline their heads . . .[27]

Aside from this theorizing about the use of movement as a tool for physiognomical analysis, the most striking feature of the *Théorie* is that of Balzac's *observateur* figure. The reader seems to be addressed as a sort of would-be *observateur,* who is learning the science from an omniscient theoretician. This theoretician, author of the text, uses the *observateur* as the ideal to which all should aspire. Indeed, the *Théorie* is greatly concerned with defining this vague and idealized figure:

> L'observateur est incontestablement homme de génie au premier chef. Toutes les inventions humaines découlent d'une observation analytique dans laquelle l'esprit procède avec une incroyable rapidité d'aperçus. Gall, Lavater, Mesmer . . . sont tous des observateurs. Tous vont de l'effet à la cause, alors que les autres hommes ne voient ni cause ni effet.
> Mais ces sublimes oiseaux de proie qui, tout en s'élevant à de hautes

régions, possèdent le don de voir clair dans les choses d'ici-bas, qui peuvent tout à la fois abstraire et spécialiser . . . ont . . . une mission purement
métaphysique. La nature et la force de leur génie les contraint à reproduire dans leurs œuvres leurs propres qualités. Ils sont emportés par le
vol audacieux de leur génie, et par leur ardente recherche du vrai, vers
les formules les plus simples. Ils observent, jugent et laissent des prin-
cipes que les hommes minutieux prouvent, expliquent et commentent.
(145–46)

This veritable cult of the *observateur* is clearly another debt to
Lavater: the genius of the "homme de science" is celebrated, fetishized, and given a place of central importance. Furthermore, as in
Lavater, the genius attributed to this mythical figure is something
which seems to come from outside of him, and over which he has little
control. Balzac explains to us that this is a force which compels the
observateur to subjectivity in his writings ("leur génie les contraint à
reproduire dans leurs œuvres leurs propres qualités"), and thus, one
would assume, to disregard other systems of rhetoric or even logic.
It is precisely because the true *observateur* is so imbued with genius
that he expresses himself in axiomatic, enigmatic discourse. Balzac
specifies as well that it is for lesser men ("les hommes minutieux") to
prove the fanciful theories produced by genius; indeed it is for them
to explicate and render these theories comprehensible. The message
is clearly that it is beneath the dignity of the *observateur* to systematize
his erratic genius into any kind of formal or conceptual coherence.

It is obvious that Balzac includes both Lavater and himself in this
lofty company of genial *observateurs*. This provides him with a perfect
intratextual defense for his Lavaterian lack of coherence as a theoretician, and serves as a backup defense of the negation of authorial
responsibility through self-parody we saw above in the *Traité*. Balzac
seems to have protected himself from any possible criticism: if we dis-
agree with his theories he can always claim to have been lampooning
them, not espousing them; if that fails, he can rely on the bottom-line
defense put forth in the *Théorie*—that as a true *observateur* in the tradition of Lavater, he is in fact obliged to produce not a systematized
theory but pearls of axiomatic "truth" to be puzzled over by those less
gifted than he.

There seems to be a definite progression from the *Traité de la vie
élégante* to the *Théorie de la démarche*[28] with the revelation of the system
of authorial self-justification outlined above. Although the *Théorie* is
filled with as much ambiguous parody as the *Traité*, there are also a
few instances in the text of what would appear to be serious reflection
on its topic and on the very process of observing and theorizing itself.

If we are tempted to dismiss all of Balzac's "physiological" writings as mere parody because of their refusal to adhere to conventions of form and tone, we need only read the passage from the *Théorie* cited above to realize that Balzac did not deem a playful tone and fragmented textual structure incompatible with serious reflection on a semiotics of the body.[29] Using Lavater as his prototype, Balzac creates a text whose parodical surface belies its conceptual agenda, a text which thus requires a "physiognomical" reading. The most important lesson given by the text in the science of observation and interpretation is the reading demanded by the text itself.[30]

If Balzac's attempts at presenting his "physiological" principles through a mixed medium of theory and parody leave their readers perplexed, it is perhaps because the texts reflect an uncertainty about the very writing of theory (hence the use of parody as a screen). One remarkable feature of these strange texts is the frequent interruption of theoretical observation by narrative: there would seem to be an irresistible urge to illustrate by anecdote, by narrative example. In fact, in keeping with the loose structuring of the entire text, instances of narrative are not necessarily identified as illustrations of particular points of axiomatic observation. Given the insurmountable drive to narrate, it is not surprising that Balzac's "physiological" principles should find a happier forum in a narrative mise-en-scène.

La Vieille Fille

La Vieille Fille, written in 1836, is the most explicitly physiognomical of all Balzac's novels. Indeed, it was the publication of *La Vieille Fille* that provoked the periodical *Le Charivari* to publish a stingingly sarcastic indictment of Balzac as the inventor of the "méthode physiognomique" in the novel:

> Cette science consiste à deviner le cœur humain, ses qualités, ses aptitudes de toute sorte d'après certains signes extérieurs. Ainsi M. de Balzac vous dira: tel homme commence sa promenade du pied gauche?—preuve d'une grande disposition pour les langues du Nord. Telle femme a l'habitude de porter ses cheveux en tire-bouchon?—vous pouvez être convaincu qu'elle est d'une certaine force sur la confiture d'abricot.[31]

Although the above is a grossly distorted interpretation of Balzac's approach to physiognomical concerns, it is nonetheless true that physiognomy can indeed be said to be the central theme of *La Vieille Fille*. In fact, physiognomy serves as more than a theme; it serves as the moral of the story. *La Vieille Fille* is above all a cautionary tale: the story of a failed physiognomical reading and its tragic consequences. Ironically, this failed reading on the level of the plot is echoed by our own necessarily failed (or at least highly imperfect) reading of the physiognomical themes of the novel itself.

Mademoiselle Cormon, the title character of the novel, is a forty-year-old heiress living in the small provincial city of Alençon and avidly seeking a husband. Not only is she eager to avoid the fate of the *vieille fille* and provide herself with companionship, she is also haunted by both persistent sexual desires and the desire to have a child. Therefore, she needs to find a man who will be a husband to her in more than name alone. Although she is not aware of each of these possibilities, we know that she has three feasible marital prospects: the chevalier de Valois, a sixty-year-old courtly aristocrat; du Bousquier, a forty-year-old bourgeois and former Bonapartiste; and Athanase Granson, a twenty-year-old Romantic idealist who alone loves her. Because of her complete inability to decipher various physiognomical signs, Mlle. Cormon chooses du Bousquier, seemingly the most virile candidate. The choice is a tragic one, as du Bousquier is in fact impotent, ambitious, and abusive. None of her desires for a "real" marriage are fulfilled, and, paradoxically, she ends up as a sort of married *vieille fille*. In the meantime, her blindness to Athanase's love for her eventually leads to his suicide.

Balzac states the moral of this tragicomic tale in no uncertain terms, and thus defines the very real extratextual agenda of his narrative:

> Ne démontre-t-elle [cette histoire] pas la nécessité d'un enseignement nouveau? N'invoque-t-elle pas, de la sollicitude si éclairée des ministres de l'instruction publique, la création de chaires d'anthropologie, science dans laquelle l'Allemagne nous devance? . . . Si mademoiselle Cormon eût été lettrée, s'il eût existé dans le département de l'Orne un professeur d'anthropologie, enfin si elle avait lu l'Arioste, les effroyables malheurs de sa vie conjugale eussent-ils jamais eu lieu? Elle aurait peut-être recherché pourquoi le poète italien nous montre Angélique préférant Médor, qui était un blond chevalier de Valois, à Roland dont la jument était morte et qui ne savait que se mettre en fureur. (226)

It should be made clear here that the "anthropologie" to which Balzac refers is a study of man more synonymous with physiognomy than with the discipline we know as anthropology today. It is also of primary importance to note that Balzac not only calls for such a discipline to be institutionalized (with "chaires"), but also for our learning to interpret the physiognomical lessons of literature itself. In trying to emphasize the need for physiognomical education, he laments not that we haven't read the treatises of Aristotle or Lavater, but rather that we haven't properly interpreted *Orlando furioso,* a work of fictional narrative. Balzac would seem to suggest that fictional narrative is indeed as appropriate a forum for illustrating the necessary lessons of physiognomical observation as physiognomical theory per se. I would argue that Balzac is indirectly validating the agenda of his own work and that of other explicit literary inscriptions of physiognomical thought here, through the double implications of the notion of literacy in the passage cited above. It is in keeping with this notion of literary discourse as a potential locus of physiognomical truth that Balzac would seem to attempt to create, in *La Vieille Fille,* a fictional narrative whose lesson cannot possibly be misconstrued, a work which contains both a narrative and its interpretation. The acute—and, one assumes, unintentional—irony of the work is of course, as I shall discuss below, that the novel ends up as more of a mystification than an illumination of physiognomical principles and practice.

Perhaps the clearest interpretation of the novel is that which is suggested by the sociopolitical identities of the two suitors actively competing for Mlle. Cormon's hand, Valois and du Bousquier. *La Vieille Fille* can indeed be read quite convincingly as a political allegory in which Valois represents the *ancien régime* and the aristocracy, du Bousquier "liberalism" in various forms and the bourgeoisie. In this reading, Mlle. Cormon would represent France itself (or more precisely, the French provinces), wavering between the two contradictory ideologies.[32] Indeed, one of the tragic side effects of Mlle. Cormon's choice of husband is that his wielding of liberal political power alienates her forever from cherished ties with the aristocracy: in marrying the paradigmatic bourgeois, she "condemns" herself to imprisonment within the bourgeoisie and alienation from the local aristocracy. Her marriage to Valois, on the other hand, would have had quite the opposite social consequence.

As far as sociopolitical physiognomy is concerned, as we saw above in the *Traité,* clothes provide the most accurate indices. In the most extensive study on the signification of clothes in Balzac, "Balzac et la Vestiognomie," Jeanne Reboul outlines the directly contrasting sartorial elements of the two suitors and their sociopolitical implications:

> Le Chevalier de Valois porte un costume de transition nuancé d'ancienne monarchie et d'élégance anglaise; Du Bousquier garde celui du spéculateur de la Révolution. Chaque détail du costume de Du Bousquier répond à un détail de celui du Chevalier, les bottes à revers aux sou-

liers vernis à boucles d'or, la culotte courte en drap côtelé a la culotte demi-juste en peau de soie, le gilet à la Robespierre à celui qui monte jusqu'au cou, l'habit bleu à l'habit marron. Ce n'est plus le contraste d'un "jadis" et d'un "aujourd'hui," mais celui de deux "jadis" dont l'un est plus ancien que l'autre, à peine démodés d'ailleurs et, plutôt qu' "historiques," devenus représentatifs de toute une manière de vivre, de toute une philosophie.[33]

Clothes, we can see, therefore serve as the first set of physiognomical indices and as the basis for a reading of the novel as political allegory. The mirroring of sartorial detail Reboul remarks on above is carried on at the level of form as well as content: the portraits of the two men are structurally quite similar. Each of the characters is introduced through a lengthy portrait which includes details about his age, physical traits, biography, and political identity, before rejoining the "present" of the narrative. This textual parallelism in the presentation of the two characters is not without significance. The men seem to function as an inseparable pair, each of whom depends on contrast with the other for his meaning. Bernard Vannier makes the astute analogy between such a pair of "personnages contrastes" and pairs of signs in Saussurean linguistics whose ability to signify is created by their difference:

Dans la mesure où "l'essentiel de la langue . . . est étranger au caractère phonique du signe linguistique," où elle est système oppositionnel et différentiel, la signification ne peut naître que dans le contraste et l'opposition des signes. Comme le système phonologique, le discours littéraire est pris dans le jeu du système sémiotique. Pour nous limiter aux portraits physiques, et sans prétendre retracer le déroulement du travail du romancier ou la conscience qu'il en aurait, le roman pose arbitrairement certains signes qui ne peuvent signifier qu'en s'opposant, et en se combinant, à d'autres.[34]

If the two characters function in this differential relationship, it is precisely because the differences between them are of such primary importance to the narrative. As informative as it is to use the vestiognomical descriptions of the characters as the basis of a political reading, it is less important to my project than the interpretation of their bodies as it is performed within the novel itself by Mlle. Cormon. The differences revealed by their bodies and her inability to recognize them form, after all, the basis of the plot.

Balzac portrays Mlle. Cormon as a good-hearted but dull-witted person who clearly lacks the powers of perception she desperately

needs. Indeed, it may be her very innocence and Christian virtue
which render her stupid:

> La dévotion cause une ophtalmie morale. Par une grâce providentielle,
> elle ôte aux âmes en route pour l'éternité la vue de beaucoup de petites
> choses terrestres. En un mot, les dévotes sont stupides sur beaucoup de
> points . . . [Donc] Mademoiselle Cormon pêchait aux yeux du monde
> par la divine ignorance des vierges. Elle n'était point observatrice, et sa
> conduite avec ses prétendus le prouvait assez. En ce moment même, une
> jeune fille de seize ans, qui n'aurait pas encore ouvert un seul roman,
> aurait lu cent chapitres d'amour dans les regards d'Athanase; tandis que
> Mademoiselle Cormon n'y voyait rien . . . Une femme de sentiment peut
> être une grande sotte. Enfin, une dévote peut avoir une âme sublime,
> et ne pas reconnaître les sons que rend une belle âme à ses côtés. Les
> caprices produits par les infirmités physiques se rencontrent également
> dans l'ordre moral. (100–102)

Although Mlle. Cormon's lack of perception may to some degree be
the result of her devotion and innocence, Balzac is also rather ironic
about such an explanation in the passage above. He characterizes her
shortsightedness as the "divine ignorance des vierges," but tells us
two sentences later that the most innocent sixteen-year-old would
have been able to recognize what Mlle. Cormon failed to see. Balzac
notes the seeming contradiction of an inability to read others coupled
with an inherently sensitive nature. Indeed, he likens this contradiction to a grotesquery as is sometimes produced by physical illness
("les caprices produits par les infirmités physiques").

As for Mlle. Cormon's innocence, she is a virgin, but not unac-
quainted with her own sexual desires:

> En ce moment, après avoir pendant longtemps combattu pour mettre
> dans sa vie les intérêts qui font toute la femme, et néanmoins forcée
> d'être fille, elle se fortifiait dans sa vertu par les pratiques religieuses les
> plus sévères. Elle avait eu recours à la religion, grande consolatrice des
> virginités bien gardées! Un confesseur dirigeait assez niaisement depuis
> trois ans Mademoiselle Cormon dans la voie des macérations et lui re-
> commandait l'usage de la discipline, qui, s'il faut en croire la médecine
> moderne, produit un effet contraire à celui qu'en attendait ce petit prêtre
> dont les connaissances hygiéniques n'étaient pas très étendues. (92)

Mlle. Cormon is physically tormented by frustrated sexual desire, and
this repression which leads to a veritable monomania to marry is,
perhaps more than her pious chastity, the cause of her blindness.[35]

Her sexual desire is of the greatest pertinence to her choice of husband, as obviously she wants to choose the man who is most likely to satisfy and impregnate her. While interpreting the suitors as political allegories provides us with a fascinating and pertinent reading of the novel, it is one which reaches beyond the perimeters of the text itself. If we read on the level of plot, however, we see that Mlle. Cormon's choice is less political than sexual: indeed, it is because she attempts (and fails) to read the sexual and refuses to read the political that she is condemned to live with a man who fails her both sexually and politically.

Mlle. Cormon's seemingly congenital lack of physiognomical perception is made clear by the many corporeal clues referring to the sexuality of the suitors Balzac provides. The chevalier de Valois is distinctly marked as an experienced, even prodigious lover, by both his physiognomy and his manner:

> Si son visage offrait quelques rides, si ses cheveux étaient argentés, un observateur instruit y aurait vu les stigmates du plaisir. (9)

> Le chevalier de Valois avait d'ailleurs la vertu de ne pas répéter ses bons mots personnels et de ne jamais parler de ses amours, mais ses grâces et ses sourires commettaient de délicieuses indiscretions. (7)

> Chez le coquet chevalier tout révélait les mœurs de l'homme à femmes ("ladie's man" [sic]): il etait si minutieux dans ses ablutions que ses joues faisaient plaisir à voir . . . le chevalier exhalait comme un parfum de jeunesse qui rafraîchissait son aire . . . Enfin, sans son nez magistral et superlatif, il eût été poupin. (10)

> Seulement, cet Adonis n'avait rien de mâle dans son air . . . (12)

In spite of the fact that "tout révélait" in the person of the chevalier and that an "observateur instruit" would have recognized the marks of the chevalier's active sexuality in his face, Mlle. Cormon does not perceive any of this and is in fact a sort of anti-"observateur instruit." Instead of breaking the code of physiognomical and pathognomical signs, she relies on the most superficial appearances to judge sexuality: she does not see that his "air" and apparent respect for propriety are but a conventional cover for his active libido. In mistaking "air" for reality, Mlle. Cormon ignores the imperative to interpret the subtler and more reliable signs furnished by the chevalier's body and thus comes to precisely the wrong conclusion:

> Si elle voyait un gentilhomme en lui, la fille ne voyait pas de mari. L'indifférence affectée par le chevalier en fait de mariage, et surtout la prétendue pureté de ses mœurs dans une maison pleine de grisettes, faisaient un tort énorme à monsieur de Valois, contrairement à ses pré-visions . . . Sans qu'elle s'en doutât, les pensées de Mlle. Cormon sur le trop sage chevalier pouvaient se traduire par ce mot: "Quel dommage qu'il ne soit pas un peu libertin!" (122)

The problem is not in the signs being given, both voluntarily and in-voluntarily, physiognomically and socially, by the chevalier; it is in their reception.

Du Bousquier furnishes clues as well: he has "le nez aplati" and a "regard fin . . . mais un peu éteint." Just as Mlle. Cormon failed to recognize the chevalier's "nez prodigieux" as the Lavaterian sign of an active sexuality, so conversely she fails to recognize the phallic symbol-ism of du Bousquier's "nez aplati." The chevalier's lively countenance is contrasted by du Bousquier's "regard éteint." However, as with the chevalier, Mlle. Cormon's reading does not transcend the superficial "air" and she mistakes the general "virility" of his appearance as a sign of sexual potency:

> Ses mains, enrichies de petits bouquets de poils à chaque phalange, of-fraient la preuve d'une riche musculature par de grosses veines bleues, saillantes. Enfin, il avait le poitrail de l'Hercule-Farnèse, et des épaules à soutenir la rente . . . Ce luxe de vie masculine était admirablement peint par un mot en usage pendant le dernier siècle, et qui se comprend à peine aujourd'hui: dans le style galant de l'autre époque, du Bousquier eût passé pour un vrai *payeur d'arrérages*. (39–40)

Du Bousquier would appear to be the ideal *payeur d'arrérages*, he who pays debts long past due, in this case he who will satisfy Mlle. Cor-mon's long-due sexual desires. Once again, the general "air" belies reality however, and only a close reading of more discreet signs would suffice to reveal his lack of sexual potency.

In each of the two suitors, there is one physiognomical feature which, read properly, would have revealed the necessary information. Both Valois and du Bousquier have voices which seem to contradict their "air," and in so doing, reveal the truth about their sexual potency, in accordance with the myth which links the male voice with the tes-ticles.[36] The tonality of the voice could have identified the "real man," had Mlle. Cormon known what to listen for:

> Pour tout dire, la voix produisait comme une anti-thèse dans la blonde délicatesse du chevalier. A moins de se ranger à l'opinion de quelques

observateurs du cœur humain, et de penser que le chevalier avait la voix de son nez, son organe vous eût surpris par des sons amples et redondants. (12)

And, in contrast:

Mais, comme chez le chevalier de Valois, il se rencontrait chez du Bousquier des symptômes qui contrastaient avec l'aspect général de la personne. Ainsi, l'ancien fournisseur n'avait pas la voix de ses muscles . . . C'était . . . une voix forte mais étouffée, de laquelle on ne peut donner une idée qu'en la comparant au bruit que fait une scie dans un bois tendre et mouillé; enfin, la voix d'un spéculateur éreinté. (40)

The imagery here is unmistakably phallic: the chevalier has a voice which corresponds to his "nez magistral" and is an "organe" surprising in its size and power; du Bousquier, on the other hand, has a voice compared to something soft and wet ("tendre," "mouillé")—the voice of fatigue and impotence ("éreinté"). Although the mythical *observateurs* so often alluded to by Balzac recognize the significance in all this (the reader is made privy to their observations through the narrator, although without any real mention of the principles informing this instance of corporeal signification), the entire raison d'être of this cautionary tale is that the title character fails to do so.[37]

Mlle. Cormon eventually makes her choice of husband through a combination of the bad judgment outlined above and sheer happenstance. Her frustration at a final thwarted attempt to win a husband other than one of her two local suitors dictates that she choose a bridegroom immediately. Taking the fact that he arrives the following day a few minutes earlier than his rival as a significant omen, Mlle. Cormon chooses the impotent and vulgar du Bousquier. Her error is symbolically but immediately revealed when du Bousquier, minutes after their having agreed to marry, accidentally lifts a luxuriant wig from his head and reveals that he is in fact bald. In accordance with the conventional folk wisdom that equates hair with virility, this is clearly symbolic of du Bousquier's inability to function sexually. The chevalier, who knows of du Bousquier's impotence, has the last cruel word:

Je vous félicite l'un et l'autre, dit le chevalier d'un air agréable, et souhaite que vous finissiez comme les contes de fées: "Ils furent très heureux et eurent beaucoup d'ENFANTS! Et il massait une prise de tabac.—Mais, monsieur, vous oubliez que . . . vous avez un faux toupet, ajouta-t-il d'une voix railleuse. (180)

Ironically, and significantly, even this last bit of irony is lost on Mlle. Cormon who, as the narrator tells us, "n'était point fille à comprendre la connexité que mettait le chevalier entre son souhait et le faux toupet" (180). Mlle. Cormon's ignorance is indeed so deeply ingrained that she does not even recognize that it is being pointed out to her.

Whether a deep, resonant voice or a head of thick, lustrous hair can indeed be signifiers of male sexual potency is a much less important point than the basic notion that there were indeed specific signs that, if read properly, would have indicated the better choice for Mlle. Cormon. As readers of the novel, we are shown those clues without being informed explicitly of the physiognomical rules they represent. We are thus in a situation somewhat analogous to that of Mlle. Cormon: although, as readers of a work framed by explicit allusions to physiognomy, we are more alert to the importance of corporeal clues, we are nonetheless also (like Mlle. Cormon) ultimately forced to speculate about what and how bodies signify.

Although Mlle. Cormon as physiognomist is the most apparent feature of the narrative, her own body as the object of physiognomical speculation is to be noted as well. Mlle. Cormon's facial features, not surprisingly, reflect both her virtue and lack of intelligence:

> Son nez aquilin contrastait avec la petitesse de son front, car il est rare que cette forme de nez n'implique pas un beau front. Malgré de grosses lèvres rouges, l'indice d'une grande bonté, ce front annonçait trop peu d'idées pour que le cœur fut dirigé par l'intelligence: elle devait être bienfaisante sans grâce. (90)

Balzac paints a portrait here with typically Lavaterian vocabulary, using the coded verbs which serve to link physical traits with metaphysical ones (verbs which Ian Pickup calls "psycho-physiological bridges"): "cette forme de nez *implique*", "ce front *annonçait*", "elle *devait être*."[38] The portrait reveals of course what we know to be true of her character from her actions discussed above. Neither should it surprise us that Mlle. Cormon should be described as being less than beautiful.

There is one element in her portrait that does stand out as potentially significant, however: that of androgyny. Balzac seems to emphasize a distinct absence of some conventionally feminine physical features in the *vieille fille:*

> Les pieds de l'héritière étaient larges et plats; sa jambe . . . *ne pouvait être prise pour la jambe d'une femme.* C'était une jambe nerveuse, à petit mollet saillant et dru, *comme celui d'un matelot.*

En ce moment, aucun corset ne pouvait faire retrouver de *hanches* à la
pauvre fille . . . (89–91; emphasis mine)

In a different text, these passing allusions to corporeal androgyny
might well go unnoticed or be considered unimportant. However, in
the context of a narrative the very subject of which is the interpretation
of corporeal detail, I think it is reasonable to speculate that even
such small details may indeed suggest something significant. In keeping with the characteristic synecdoche of physiognomical thought,
a single detail (like that of the voice above) can reveal the entire
"essence" of a given character.

Furthermore, the implications are strongly reinforced by a possible
causal link between this androgyny and Mlle. Cormon's perceptual
insufficiencies: the absence of feminine physical qualities is mirrored
by an absence of the mythical "feminine intuition." This instinctive
knowledge about men on the part of women is defined by Balzac early
on in the novel:

Les femmes ont un instinct qui leur fait deviner les hommes qui les
aiment par cela seulement qu'elles portent une jupe, qui sont heureux auprès d'elles, et qui ne pensent jamais à demander sottement l'intérêt de
leur galanterie. Les femmes ont sous ce rapport le flair du chien, qui dans
une compagnie va droit à l'homme pour qui les bêtes sont sacrées. (26)

Aside from the inescapably offensive misogynist analogy between
women and dogs, what is most significant in this passage is the definition of woman by her possession of this innate knowledge. The
repeated construction "Les femmes ont . . ." does not allow for the
possibility that some women may not have the supposed instinct.
The logical implication is that those who do not exhibit this species-
defining behavior are not of the species. Since the entire narrative is
based on demonstrating that Mlle. Cormon does not have this instinct,
is it too far-fetched to assert that the entire narrative demonstrates
the fact, reflected by physiognomical clues, that Mlle. Cormon is not
a "real" woman? According to the definition above, had she been a
"real" woman, would she ever have chosen an impotent woman hater
over a young man desperately in love with her and a sexually potent,
lifelong "ladies' man"?[39]

It is indeed their respective relations to women (both sexual and
social) that distinguish most emphatically the two suitors:

Du Bousquier se faisait gloire d'appartenir à cette école de philosophes
cyniques qui ne veulent pas etre "attrapés" par les femmes, et qui les

mettent toutes dans une même classe suspecte . . . Sous ce rapport, du
Bousquier était encore la contre-partie du chevalier de Valois. (50–51)

It is precisely this all-important distinction that Mlle. Cormon fails to
recognize. Perhaps only a "real" woman is able to recognize a "real"
man, the text seems to suggest.

Mlle. Cormon's marriage, around which the entire narrative is centered, is frequently characterized as that which will transform her
into a woman, from *fille* into *femme,* with all the possible implications
of both words pertaining. However, because of her fundamental inability to decipher the physiognomical signs of "virility," she chooses
the one man who will not actualize her transition into *femme.* As one
character puts it: "Mme. du Bousquier ne serait jamais que Mlle. Cormon" (199). Fredric Jameson reads this irony in the following terms:
"The final and most unexpected reversal of all will be that in which
both success *and* failure are cancelled out, in that desolate transcen-
dence of which the conclusion of *La Vieille Fille* furnishes a triumphant
and intolerable realization (married and an old maid all at once!)."[40] In
other words, the oxymoronic classificatory term *vieille fille,* which provides the novel with a title recalling Balzac's *Physiologies,* is exchanged
for the equally oxymoronic status of married maiden. *La Vieille Fille* is
the *physiologie* of a grotesque, of a person who escapes categorization
and Balzac's essentialist definitions of gender. In her failure to satisfy
the terms of these definitions as both physiognomist and object of
physiognomy, Mlle. Cormon reveals that she is truly, inescapably an
oxymoronic being, the quintessential *vieille fille.*

Perhaps even more to the point for the purposes of this study, the
novel renders its reader equally inadequate in his/her quest for the
physiognomical enlightenment it would seem to promise. Without
any specific didactic gestures in the text, can we as readers be expected to understand the logic that links a woman's "masculine" leg
to her innate inability to read character and sexual potency in a man?
The essentialist (and ludicrous) definition of gender at work here, that
a "real" woman cannot have the leg of a sailor, but must possess hips
and intuition about men, is never clearly articulated and is far from
obvious. The novel slyly alludes to such notions, but ultimately leaves
its reader to wonder on his/her own about the code of physiognomical knowledge on which these allusions are founded. Similarly, we
are left to puzzle over the possible "scientific" bases of the relations
between men's noses, voices, hair, and genitals implied in the descriptions of the two suitors. Much physiognomical knowledge is implied
in the narrative and by the plot, but little is divulged. There seem to

exist answers to our questions, but we remain incapable of arriving at them on our own. If the reader of *La Vieille Fille* knows more than Mlle. Cormon, he/she clearly knows considerably less than the coyly winking narrator of the tale, who declines to share any specifics of his superior knowledge of the signifying body with us.

La Fille aux yeux d'or

We have seen aspects of Balzac's physiognomical thought expressed through the media of theory (albeit semiparodical theory) and fictional narrative. Before concluding, however, there is a unique text which merits at least a brief analysis precisely because it joins the two discourses in a single literary work. *La Fille aux yeux d'or* is made up of two distinct parts; the first is a sort of introduction which analyzes the "physiognomy" of Paris through the classification of its socio-economic types, the second a mysterious and aberrant tale of a sexual liaison in the city. The relation between these two parts of a single text is enigmatic. One might well have the initial impression that they are simply unrelated pieces of writing given a superficial link by an author eager to join sociological speculation with fictional narrative. There is, however, a dialogue between the two: although the sociological introduction fails to illuminate the narrative (as Balzac would have it), their relation is nonetheless the most suggestive aspect of the work, and the one most pertinent to our study.

The first part of *La Fille aux yeux d'or*, which Balzac published separately under the title *Physionomies parisiennes* prior to publishing the novella itself,[41] is an attempt to classify the various strata (referred to in the title by the anthropomorphic "physionomies") of Parisian society according to the principles of money and pleasure. Each class is constituted according to its relation to, and pursuit of, these fundamental social phenomena. As was the case above in the *Physiologies*, all of the sociological commentary is offered by a neutral, objective "observateur" and is modeled after the discourse of zoological classification. As Michel Nathan tells us, this system of classification in *La Fille aux yeux d'or*, based on the comparison of functions common to all beings, is borrowed from Geoffroy Saint-Hilaire:

> Comment Balzac classe-t-il les Parisiens? En fonction de la manière dont toutes les activités qui épuisent leur capital d'énergie s'organisent et convergent vers l'or et le plaisir . . . Tous les animaux respirent. Cette fonction est assumée, selon les espèces, par diverses associations d'organes, ce qui peut être à l'origine d'une classification. Tous les Parisiens cher-

chent l'or et le plaisir. On peut les classer selon la forme que prend, chez chacun d'eux, cette aspiration.[42]

  From this seemingly objective bird's-eye view of Parisian society, Balzac focuses in on a detail of the big picture, the particular tale of a particular individual. The movement from type to individual, from general to particular, is reflected by the generic discursive shift from theory to narrative. The narrative part of the work recounts the chance encounter of a fashionable young dandy (Henri de Marsay) and a beautiful, mysterious young girl (Paquita Valdès). Although the girl is closely guarded by a formidable *duègne*, Marsay becomes wildly infatuated with her extreme and rare beauty, particularly that of her golden eyes.[43] Through a baroque series of machinations, Marsay and Paquita are able to consummate their passion in great secrecy from the girl's unidentified keeper. Through several odd occurrences during their tryst, Marsay figures out that Paquita's enigmatic half innocence in matters of sexuality is attributable to the fact that she is a virgin only to heterosexuality, but not, however, to homosexuality. He returns to seek revenge, only to discover that Paquita has been murdered by her lesbian lover. As he and his female rival face each other, they realize that they are half-siblings.

  Paquita, who is constantly referred to by the physiognomical appelation "la fille aux yeux d'or,"[44] is characterized not by the simple *fact* of her beauty, but rather by the rare quality of her beauty. In direct contrast to the conventional beauty of most of Balzac's heroines (conventions which date back to the twelfth century), Paquita is not a golden-haired, blue-eyed, alabaster-skinned maiden.[45] In a typically Balzacian gesture of physiognomical coherence, this rare beauty is balanced by a rare sexual status: that of semivirgin, one who is sexually experienced, but uninitiated to the most conventional sexual practice (heterosexual intercourse), to that which traditionally makes the distinction between virgin and nonvirgin. As this semivirginity is a condition for which no name exists, it comes as no surprise that Marsay finds some difficulty in divining it as the explanation for Paquita's strange combination of innocence and knowledge in sexual matters. The plot is indeed fueled by Marsay's dual desire both to possess a beauty so unlike that of his other numerous conquests and to gain access to her (sexual) "secret."[46]

  Paquita's gold eyes seem to suggest that her body itself is the unique incarnation of an essential link between *or* and *plaisir*. Gold eyes are indeed quite rare (if they can be said to exist at all); even Pierre Abraham, in his seemingly exhaustive study of hair and eye color in the

Comédie humaine,[47] fails to mention this aberrant case of pigmentation. The use of the term "yeux d'or" invokes the surreal image of eyes whose actual substance is of gold, rather than simply their color. Paquita thus echoes, in her beauty and sexuality, the *or* and *plaisir* of the sociological system outlined in the first part of the work.[48]

However, because her beauty is such a rare one, she herself defies the categorization inherent in sociological systems. In the first part of the text, Balzac identifies the prototypical faces of the various classes he describes and the ravages effected by the quest for gold and pleasure. The entire narrative of the second part of his work is conversely based on a quest for the gold and pleasure of a face which is in no way comparable to any other, which escapes and subverts both physiognomical and sociological classification.[49] Balzac makes explicit the fact that the analysis of ordinary physiognomies which constitutes the sociological part of his text serves to justify the recounting of the narrative of a rare, unclassifiable face:

> Si ce coup d'œil rapidement jeté sur la population de Paris a fait concevoir la rareté d'une figure raphaëlesque, et l'admiration passionnée qu'elle y doit inspirer à première vue, le principal intérêt de notre histoire se trouvera justifié. *Quod erat demonstrandum,* ce qui était à démontrer, s'il est permis d'appliquer les formules de la scolastique à la science des mœurs.[50]

The bizarre events recounted by the narrative, all directly or indirectly provoked by the rare face, are equally beyond being classified or ordered, as they so greatly defy comparison with the usual events of human life. The two parts of the narrative point up the contrasting functions of the two modes of discourse: theory, which by definition attempts to generalize, and narrative, which recounts the particular. Serge Gaubert expresses this opposition in *La Fille aux yeux d'or* nicely:

> Il n'est de bon roman que de l'exceptionnel: un homme, un jour, un homme singulier, un jour marqué; et le vent pousse loin des côtes basses et fort fréquentées, au hasard des événements. A l'habituel, à la loi, au portrait robot du tableau répond l'étrangeté d'une figure atypique, à la similitude la pure différence, au concentrique l'excentrique . . . Mots phares d'abord, héros-questions ensuite.[51]

It is therefore impossible to read the first part of the text as a sort of theoretical key to the second part: *La Fille aux yeux d'or* does not use narrative to illustrate theory, nor theory to illuminate narrative. Shoshana Felman points out how the ambiguous relation between the two modes is revealed by the ambiguity of the word *or* itself which serves as transition from the theoretical to the narrative part of the text:

> Constituting the ambiguous *transition* between the introductory authoritative discourse and the narrative, joining the prologue's gold and the story of the girl with the golden eyes, does the signifier, the conjunction OR, mean "thus" (logically introducing an argument in support of a thesis) or does it mean "however" (logically introducing an objection)? Through this conjunction, does the story serve to *illustrate* the prologue, as a slave would serve its master (OR = "thus"), or is the story, on the contrary, a rhetorical *subversion* of the authority (of the paternity or consciousness) of the prologue (OR = "however")? . . .
> It could be said that . . . the narrative indeed ends up subverting, to some extent at least, the "guiding light" of the prologue, the authoritative truth which was supposed to be its "proper" meaning and which it had to "demonstrate" as a self-evident principle of identity and value which . . . organizes Paris as a social universe of order . . . [52]

Indeed, the text uses narrative to illustrate the existence of certain cases, certain spheres of human being which escape the generalizing sweep of theory, both physiognomical and sociological. It becomes the function of narrative to recount that which escapes theory. These exceptional cases would seem to subvert the very agenda of Balzac's "physiology," by proving that no such system can ever express the totality of human existence (or even, in this case, of Parisian existence). Human life is such that there will always be cases which escape classification. The very possibility of a girl with golden eyes puts all totalizing statements about physiognomies into question.

The acts of seeing, reading, and interpreting are indeed problematized in various ways in *La Fille aux yeux d'or*. From the beginning of the narrative section of the text, this is suggested by the contradiction between Marsay's youthful, innocent, almost virginal appearance and his callous, world-weary, macho character:

> Quoiqu'il eût vingt-deux ans accomplis, il paraissait en avoir à peine dix-sept. Généralement, les plus difficiles de ses rivaux le regardaient comme le plus joli garçon de Paris . . . Les yeux bleus les plus amoureusement décevants . . . les cheveux noirs les plus touffus . . . un sang pur, une peau de jeune fille, un air doux et modeste, une taille fine et aristocratique, de fort belles mains. (262)

Although the above is an excellent example of physiognomical *signes trompeurs*,[53] and therefore significant, reading is more suggestively, if more metaphorically, problematized by the blindfold Marsay is obliged to wear en route to his assignations with Paquita. This literalization of the impossibility of seeing and interpreting for Marsay inversely validates the equation *voir = savoir = pouvoir*. When he is first blindfolded, Marsay struggles fiercely, as if he senses the implications of his triple loss of sight, knowledge, and power:

> Si vous voulez venir . . . il faut consentir à vous laisser bander les yeux.
> Et Christemio montra un foulard de soie blanche.
> Non! dit Henri dont la toute-puissance se révolta soudain.
> Henri était fort, il voulut se jouer du mulâtre. Lorsque la voiture partit au grand trot, il lui saisit les mains pour s'emparer de lui et pouvoir garder, en domptant son surveillant, l'exercice de ses facultés afin de savoir où il allait. (302)

Having lost the struggle, Marsay submits to a journey in literal darkness, much as his entire liaison with the girl with the golden eyes has been a journey into the figurative darkness of ignorance. However, in contrast, his second rendez-vous with Paquita comes after he has developed a credible hypothesis about her unknown life; his knowledge and power therefore more or less intact, he has little to fear, and acquiesces graciously to the blindfold ("il . . . se laissa complaisamment bander les yeux" [317]). Once Marsay has verification of his theory (Paquita's confused use of the name "Mariquita" in a moment of sensual abandon), his will to possess the beautiful woman is replaced by the will to destroy her. At the moment of his ultimate revenge, however, the bizarre and unforeseeable circumstance of his rival Mariquita's being his own sister ironically invalidates his momentarily regained power to "see." No clue could have revealed the grotesque scenario to Marsay; reading would seem to have been rendered impossible for him as well as for us.

While it is a completely valid rhetorical reading to interpret the narrative of *La Fille aux yeux d'or* as negating the urge to theoretical imperative suggested by the introduction, it is perhaps equally valid to try to read the relation in light of the larger context of Balzac's work. It has been more than sufficiently demonstrated that attempting to reconstruct authorial intention is a foolish and unsatisfying pursuit. Nonetheless, we are left with nagging questions about how we might reconcile the seeming negation of theory enacted by the narrative with the theoretical imperative Balzac expresses so prolifically elsewhere in his œuvre. I would speculate that this narrative of faces and events which escape typology points out that there are indeed always exceptions, but that these exceptions do not negate the rules. The consummate creator of systems and categories does not interpret the few inevitable exceptional cases as destroying the power of his theory, even if his urge to narrate compels him to recount the exceptions.

In spite of exceptions, Balzac would seem to believe, we still need rules and systems of ordering human existence. The violent and bloody dénouement of the story of the girl with the golden eyes suggests the danger of not being able to read a face and the life it represents; had Marsay been able to read Paquita, he would never have made love to her, and she, consequently, would never have been murdered. The message of the text, then, is not necessarily contradictory to that of *La Vieille Fille* and the various "theories" we saw above. The tragedy of Mlle. Cormon's story is her inability to read obvious clues; the tragedy of *La Fille aux yeux d'or* is that of the exceptional case which *cannot* be read, even by as savvy an *observateur* as Henri de Marsay and certainly not by a mere reader of the unpredictable and unexplained tale. One might say that the tragedy of *La Vieille Fille* is precisely that of evitability, while the tragedy of *La Fille aux yeux d'or* is that of inevitability. Neither case implies, however, that the acts of observation and interpretation are futile ones, simply that some (the narrator) are capable of performing them while others (the characters and the reader) are not.[54]

It is not certain that Balzac would have wanted to be considered a sociologist as such, given the parodical tone of many of his writings on the subject. However, it is clear that the observation, interpretation, and classification of human beings is a constant, even obsessive, theme in many of his works. As we saw in the Avant-propos, these acts provide the very foundation of his narrative project. Balzac's œuvre is perhaps at its most suggestive when it is read as posing questions about the relation between sociology and narrative, and the implications of each for the other. As is true for any novel which can be more or less classified as an *histoire de mœurs*, Balzac's fictional works reflect much about the *milieux* in which they are set. His own use of the term *physiologie* to describe an intellectual endeavor which blends physiognomical and sociological questions into a single process of observation and interpretation best characterizes his concerns. Because he uses corporeal signs as indicators of both sociological and characterological truths, his depictions of characters are always concerned with rendering the human body intelligible. It is in fact Balzac's insistence on signification and intelligibility and the ways in which they are both thematized and enacted by his texts that make the *Comédie humaine* an exceedingly rich and worthy object of literary analysis.

Balzac's novels have been criticized, as have all "realist" works, for their excessive descriptions. These passages are often read as failures of the narrative economy, superfluous digressions, or textual windowdressing. Nothing could be less accurate, however, as Balzacian description (particularly that of the "portrait" of a character) is never gratuitous, nor even *effet de réel*.[55] Because Balzacian description is always "physiological," and the physical is always revelatory of the metaphysical, the reader is continually having his understanding of the narrative itself reshaped by descriptive passages. Bernard Vannier defines this function of the portrait: "On sait que la fonction essentielle et la plus commune du portrait est d'éclairer l'action. Ce serait un document, destiné à faire voir . . ."[56] Indeed, descriptions create their own subnarrative of explication and illumination. Therefore, even as seemingly trivial a detail as the hair color of a character may play a role in a complex intratextual network of correspondence between superficial phenomena and essential truth. Descriptive details often serve as interpretative clues to what Balzac wants us to understand about his narrative and its characters.

The function of description in the Balzacian text reflects the fundamental rhetorical agenda of the text itself, which is to deny the arbitrary and ascribe an absolute power of signification to all persons, objects, and phenomena.[57] Peter Brooks clearly defines this agenda in the following manner:

> Here we touch on the core of the Balzacian project and aesthetic: to make the plane of representation imply, suggest, open onto the world of spirit as much as can possibly be managed; to make the vehicles of representation evocative of significant tenors. Meaning is ever conceived as latent . . . Balzacian description is regularly made to appear the very process of investing meaning in the world, demonstrating how surface can be made to intersect with signification . . . Everything in the real— façades, furniture, clothing, posture, gesture—must become sign.[58]

"Physiological" description is the very juncture between the planes of representation and signification,[59] and is thus the rhetorical and conceptual nexus of Balzac's texts. The phenomenon is further complicated, however, by the fact that the process is often one of creating, rather than uncovering, an unquestionable meaning. Balzac manufactures meaning where it need not necessarily exist (in a human face, for example). As Brooks says above, the plane of representation is

made to imply, description is *"made* to appear the very process of in-
vesting meaning in the world," and "surface can be *made* to intersect
with signification . . ." (emphasis mine). If we choose then not to accept Balzac's "world of hypersignificant signs"[60] as an absolute truth
of human existence (as he would have it) and define it rather as his
own authorial construct, we must also examine the manner in which
he imposes this process of signification on the reader. In the Balzacian
text, meaning is indeed imposed rather than implied.

The system of hypersignification is communicated through the
medium of an omniscient, impersonal narrator. This vague but authoritative voice informs the reader of the various signifieds represented by the various signifiers in the text; in short, it reads for him/
her. The Balzacian novel thus becomes a totality, comprising not only
narrative and description, but even exposition of the act of semiotic
interpretation which ties the two modes together.

In *The Act of Reading: A Theory of Aesthetic Response,* Wolfgang Iser
defines the literary work as a performative construct which comes into
being through the response of its reader rather than existing as a transparent medium.[61] He defines "indeterminacy" in a narrative text as a
prerequisite for interaction between the text and its reader and thus
a "basic constituent of communication."[62] For Iser, the act of reading
involves the working out of a solution to these indeterminacies.[63] It is
clear that, at least according to a definition of the act of reading such
as Iser's, Balzac's texts do not freely lend themselves to be "read," as
there are few such indeterminacies to be resolved. Through the intratextual explication of a universe of hypersignification, the text has in
effect already at least attempted to "read" itself.

As a hermetic and saturated piece of writing, the Balzacian text
seeks to preclude any true act of interpretation on the part of its
reader. Nowhere is this more clearly demonstrated than in the constant references to codes and systems of knowledge to which only the
narrator (necessarily) has access. Balzac's *physiologie,* a unique hodge-
podge of principles and methods borrowed from zoology, sociology,
and physiognomy, is a code of which only he has complete understanding. For the most part, readers of texts founded on the code must
therefore passively observe a process of interpretation, without being
able to effect their own analysis. The process of interpretation must
ultimately, at least in principle, be left up to the narrator himself and
to the mythical but omnipresent *observateurs du cœur humain.*

Just as the reader of the Balzacian narrative is rendered passive and
unquestioning before the omniscience of the narrating *observateur,* the
status of the truth as known by the *observateur* is never doubted. In

spite of their often playful and even ironic tone, Balzac's texts share a fundamental subtext of subscription to the meaning they create. We, as savvy poststructuralist readers, can argue that such meaning is always manufactured by the text itself and is therefore no real "truth" at all. However, the texts themselves posit, with a great degree of certainty, an adherence to various systems of meaning, not least of which are those of physiognomical thought and gender essentialism. The reader is denied access to knowledge crucial to a complete understanding of the principles thematized in the novel, but never are the principles themselves genuinely put into question.

One is tempted to consider our skepticism vis-à-vis the essential truths espoused in Balzac as merely a function of our historical distance from the texts. The difference, however, is both far more interesting and far less historical than that. Gautier, a contemporary of Balzac, demonstrates a strikingly more modern view of the "truth" of gender, sexuality, and signification than that we have seen in the preceding chapter. Rather than echo his friend Balzac's ultimate belief in such systems, Gautier prefigures our own late-twentieth-century skepticism about the reality of any essential corporeal semiotics. Like Balzac's *La Vieille Fille* and *La Fille aux yeux d'or*, Gautier's *Mademoiselle de Maupin* (1836) tells the story of people and situations which are far from typical and which thus test the limits of conventional definitions of the meaning of the body. Unlike Balzac's narratives, however, the ultimate consequence of Gautier's novel is to refute, rather than reaffirm, essentialist systems of corporeal signification. Gautier's text is one which asks questions to which it does not provide answers, and thereby subverts the very notions Balzac seeks to maintain.

Chapter Five

Gautier and *la beauté de l'ambigu*

Comme toute chose au monde, la beauté morale ou physique veut être
étudiée, et ne se laisse pas pénétrer tout d'abord.

—Gautier

Mademoiselle de Maupin, first published in two volumes in 1835 and
1836, is the most renowned of Théophile Gautier's works of prose. The
critical reception accorded this strange quasi-epistolary tale of sexual
and moral ambiguity has been, from its publication until relatively recently, rather confused.[1] It should perhaps come as no surprise that
critics have found *Mademoiselle de Maupin* an unusually difficult text to
interpret: its narrative structure is something of a patchwork, its content at times abstract and dense, its plot far-fetched, and its tone ambiguous. Many critics have simply sidestepped the question of textual
analysis by seeking recourse in biographical interpretation.[2] Others,
such as Anne Bouchard, have noted the complexity and difficulty of
the work and of the "presence" of its author as part of their analysis:
"Il s'agit bien là d'une oeuvre exceptionnelle, dont la plus belle réus-
site est, peut-être, pour son auteur, d'avoir su, sous tant de masques
divers, se projeter dans tant de miroirs, partout présent, pourtant
insaisissable."[3] Indeed, Bouchard quotes René Jasinski's characterization of the novel as one which seems to "defy analysis."[4]

Fortunately for the purposes of this study and in spite of the elusive nature of the text, recent critics and students have provided me
with insightful and illuminating readings on which to base my own
interpretation of the novel; my debt to their work will no doubt be
obvious.[5]

The great diversity of approaches to the novel notwithstanding,
most critics seem to agree on the centrality and fundamental importance of the theme of the body in *Mademoiselle de Maupin*. It can be
said with some certainty that no other novel, even in the corporeally
obsessed nineteenth century, more explicitly thematizes the human
body, both as a self-referential entity and in its complex and ambiguous relation to the domains of the soul and the mind. It is therefore

a text which emphasizes the most fundamental questions of physiognomy: that of the link between the physical and metaphysical realms, and more important, that of the possibility of the legible body, the body as signifier.

Clearly, the central problem at issue in *Mademoiselle de Maupin*, a story of androgyny and desire very loosely based on that of the infamous seventeenth-century actress, transvestite, and bisexual of the same name, is that of reading sexual identity. The ambiguity created by the eponymous character's transvestism forces d'Albert, the male narrator of much of the novel, to read not only his/her sexual identity, but his/her gender identity as well. This unusual physiognomical exercise points up not only physiognomical concerns per se, but also the very question of the possibility of defining gender. Is sexual identity a purely anatomical condition, therefore always more or less legible through scrutiny of the body? Or is it determined by the object of one's sexual desire—in other words, is a woman who desires another woman a "man," as is suggested at several points in the text? Can a person have the body of one sex and the mind of the other? That is to say, are sex and gender necessarily identical? Gautier's text raises these and other questions, not in order to provide glib or facile answers, but in a complex and provocative manner, through both a compelling narrative and lengthy soul-searching on the part of his two narrators. These are questions which plumb the depths of our understanding of the notion of gender itself and thus of the relation between the physical and the nonphysical. They are indeed questions still being asked today, albeit in somewhat more sophisticated terms.[6]

A question less often addressed, but equally compelling, is that of the body as esthetic object. Gautier's cult of beauty is both well-known and well documented.[7] In *Mademoiselle de Maupin*, d'Albert finds himself in the unusual position of admiring, and desiring, a body of indeterminate sex. If a body can be treated as a "purely" esthetic object, why should sex *have* to be read? Is beauty necessarily gendered? If so, what is the relation between esthetics and erotics? What are the ethical implications of reducing a human being (of either sex) to an esthetic object? As we shall see, in direct contrast to most of the other literary works in this study, *Mademoiselle de Maupin* is a text far richer in questions than in answers.

The Preface to *Mademoiselle de Maupin*

The relation between esthetics and morality (or lack thereof) is a point which is central to Gautier's oeuvre. Indeed, perhaps the best known and most influential discussion of the subject in the nineteenth century is his preface to *Mademoiselle de Maupin*. This infamous diatribe has two fairly simple theses concerning beauty: first, that utility and beauty are mutually exclusive categories ("Rien de ce qui est beau n'est indispensable à la vie"; "Il n'y a de vraiment beau que ce qui ne peut servir à rien"),[8] and second, that questions of beauty are necessarily amoral. In short, Gautier tells us, there can be no relation between esthetic and moral concerns other than that of mutual exclusion. He illustrates his point about this nonrelation, and about the superiority of esthetics, in the following rather tongue-in-cheek allegory of virtue as a grandmother and vice as a beautiful *coquette:*

> La vertu est assurément quelque chose de fort respectable, et nous n'avons pas envie de lui manquer, Dieu nous en préserve! La bonne et digne femme!—Nous trouvons que ses yeux ont assez de brilliant à travers leurs bésicles, que son bas n'est pas trop mal tiré, qu'elle prend son tabac dans sa boîte d'or avec toute la grâce imaginable, que son petit chien fait la révérence comme un maître à danser.—Nous trouvons tout cela.—Nous conviendrons même que pour son âge elle n'est pas trop mal en point, et qu'elle porte ses années on ne peut mieux.—C'est une grand'mère très agréable, mais c'est une grand'mère . . . Il me semble naturel de lui préférer, surtout quand on a vingt ans, quelque petite immoralité bien pimpante, bien coquette, bien bonne fille, les cheveux un peu défrisés, la jupe plutôt courte que longue, le pied et l'oeil agaçants, la joue légèrement allumée, le rire à la bouche et le coeur sur la main. (1)

There are several things one might note in Gautier's allegory: first, the fact that his choice of the image of an old woman with decidedly dated accessories (the *tabatière,* the lapdog) to represent virtue is highly suggestive in the literary-historical context in which the preface was written;[9] second, the use of feminine allegories of vice and virtue prefigures an objectification of women we shall see later in the novel itself, as well as at various other moments in the rest of the preface.[10]

Of most immediate interest to us, however, is the most basic and obvious point, which is that the esthetic and moral domains cannot be used to interpret each other. Gautier does not tell us that there is no such thing as a moral domain; he suggests (albeit with great irony) that vice and virtue may indeed exist. However, they are not qualities which can be evaluated esthetically. Clearly, the converse is equally true: esthetic qualities cannot be evaluated in moral terms. Janet Sadoff quotes Zola's *Documents littéraires* on the amorality of Gautier's esthetic concerns:

> Au fond la vérité ne le touchait guère. Il vivait pour le monde extérieur. Il s'en tenait à la draperie, il la trouvait belle ou laide, et le disait. Ce

grammarien, ce rhétoricien, ce peintre n'était fatalement remué que par les questions de grammaire, de rhétorique et de peinture . . .

Zola goes on to contrast Gautier's concern for physical detail for its own sake with the naturalist concern (the descendant, as has often been said, of Balzacian principles of description) for physical detail as an index to psycho-sociological (and, by extension, moral) truths:

> Le voilà encore avec son unique souci du monde tangible; toujours peintre, jamais observateur ni analyste. Le plus curieux est qu'il se rencontre ici sur un point avec les écrivains naturalistes, qui ont, eux aussi, le plus grand respect pour les milieux; seulement, eux ne les étudient soigneusement que parce qu'ils complètent et déterminent l'homme, tandis que Théophile Gautier les veut pour eux-mêmes, en dehors de l'homme.[11]

Although Gautier's specific concern in his preface is esthetics in the sense of artistic (and especially literary) creation, the absolute separation from morality is equally valid for esthetics in the more general sense of beauty in various forms, especially that of the human body.[12] Indeed, the centrality of the human body as esthetic object in this discussion could not be more clearly emphasized than in Gautier's heavy-handed use of anthropomorphic imagery.

There are other instances in the preface where Gautier emphasizes the human body as a sort of counterindex to morality, if only to show that the immorality (or, as he would have it, amorality) of the literature of his generation has such venerable predecessors as Molière:

> Que voit-on dans les comédies du grand Molière? La sainte institution du mariage . . . bafouée et tournée en ridicule à chaque scène.
>
> Le mari est vieux et laid et cacochyme; il met sa perruque de travers; son habit n'est plus à la mode; il a une canne à bec-de-corbin, le nez barbouillé de tabac, les jambes courtes, l'abdomen gros comme un budget . . .
>
> L'adultère est toujours jeune, beau, bien fait . . . il se gratte l'oreille avec l'ongle rose de son petit doigt coquettement écarquillé; il peigne avec son peigne d'écaille sa belle chevelure blondine, et rajuste ses canons qui sont de grand volume. (7)

Although it would be absurd to suggest that the central thesis of the preface to *Mademoiselle de Maupin* is one which is specifically concerned with the human body, Gautier does seem, to a student of physiognomy at least, to espouse a sort of anti-Lavaterian philosophy: whereas Lavater postulated that Jesus Christ, as the most virtuous

man who ever lived, must logically have also been the most beautiful,[13] Gautier tells us in no uncertain terms that beauty does not
(and indeed cannot) represent virtue, but merely represents itself,
in a purely esthetic domain irrelevant to moral value judgments. All
that can be said about a beautiful object (including, and especially, a
human body) is that it is beautiful. For Gautier, further commentary
is neither necessary nor possible.[14]

Furthermore, to emphasize the priority of the physical/esthetic over
the metaphysical/moral, Gautier mocks the notion of man's moral
perfectibility. He tells his reader, in an exaggerated and ironic tone
typical of the preface, that only when the human body will have been
perfected, will man have achieved real progress:

> Mon Dieu! que c'est une sotte chose que cette prétendue perfectibilité du
> genre humain dont on nous rebat les oreilles! . . . Quand on sera par-
> venu à donner un estomac double à l'homme, de façon à ce qu'il puisse
> ruminer comme un boeuf, des yeux de l'autre côté de la tête, afin qu'il
> puisse voir, comme Janus, ceux qui lui tirent la langue par derrière . . .
> quand on lui aura créé un nouvel organe, à la bonne heure: le mot *per-
> fectibilité* commencera à signifier quelque chose. Depuis tous ces beaux
> perfectionnements, qu'a-t-on fait qu'on ne fît aussi bien et mieux avant
> le déluge? (25)

The fundamental message of the preface to *Mademoiselle de Maupin*
is not only the explicit polemic against hypocritical, moralistic, anti-
Romantic critics and journalists. It is also the absolute separation of
moral questions from esthetic questions (this is, of course, the basis of
the philosophy of *l'art pour l'art*), and the novel gesture of valuing esthetic over moral criteria. As we shall see below, Gautier enlarges and
problematizes these very issues in the context of a fictional narrative
in the novel itself.[15]

Mademoiselle de Maupin

Perhaps the first thing to note, in passing, about *Mademoiselle de Maupin* is that it is characterized, as has frequently been remarked, by
a rather baroque and not entirely coherent narrative.[16] Indeed, even
Gautier's contemporary admirer Victor Hugo felt obliged to justify the
rather sloppy treatment of plot in the novel:

> *Mademoiselle de Maupin* est un livre qu'il faut lire, et surtout qu'il faut
> relire. Qui le lit peut en être mécontent, qui le relit en est charmé. A la
> première lecture en effet, ce que saisissent en général les intelligences

superficielles, c'est l'aventure, lévénement, l'anecdote, la machine, chose importante et sérieuse, à notre avis, mais que M. Gautier néglige et dédaigne . . . Ce qui apparaît à la seconde lecture, ce sont les qualités qui font l'exquise valeur du livre de M. Gautier, c'est le style charmant, c'est l'exécution parfaite, c'est l'abondance des idées, des images, des sentiments, plus amusants pour les esprits délicats que l'abondance des événements . . . [17]

A much more recent critic, Carmen Fernandez-Sanchez, also proposes (albeit with a very different vocabulary) a nonnarrative reading of the novel:

Quelque soit la structure du récit, une lecture symbolique des personnages, de l'espace et du temps de *Mademoiselle de Maupin* révèle l'insuffisance d'une lecture prosaïque qui ne tiendrait compte que du sens horizontal de l'intrigue. Lu verticalement, en recherchant le parallélisme des signifiés, ce roman n'est plus un roman irrégulier ou imparfait mais un récit à placer dans le genre des récits poétiques en prose.[18]

Because, however, my own reading of the novel incorporates narrative concerns other than those expressed by the two critics above, and precisely because the story is relatively complex, it would be perhaps useful to outline here, however briefly, the plot(s) of the novel.

D'Albert, the initial narrator, is a young would-be poet, stricken with a rather extreme case of the Romantic *mal du siècle*.[19] The disease manifests itself through the usual symptoms (ennui, listlessness, general dissatisfaction with life), as well as through some rather more particularly esthetic ones (longing for an ideally beautiful woman, desire to be beautiful himself). In order to distract himself from his *mal*, he takes a mistress who is beautiful enough to satisfy his body, but not beautiful enough to satisfy his "soul's" craving for ideal beauty. While sojourning in the country with this mistress, Rosette, he meets a remarkably handsome young *cavalier*, Théodore, whose beauty corresponds to d'Albert's preconceived ideal in every way but one: that of his (apparent) sex. His attraction is nonetheless so great that he finds himself falling in love with the young man, much to his alarm and distress.

The narration next becomes that of a young noblewoman, Madelaine de Maupin (the variant spelling of Mlle. de Maupin's given name is that used by Boschot in his definitive 1966 edition of the novel and thus, presumably, by Gautier himself; other editors have chosen to "correct" it). Before giving herself to a man, Madelaine wants to see the true "face" of men among themselves, as opposed to the insincere

posturing in which they engage in the presence of women. In order
to achieve this ambitious goal, she takes the unusual step of roaming
the countryside disguised as a young man. Her travels take her to a
château where she meets Rosette, a young widow who falls in love
with her and tries valiantly to seduce her, not knowing her true sex.
Later, she encounters Rosette again, this time in the company of her
lover, d'Albert, who in turn falls in love with her as Théodore.

As an amusement, the guests at the château decide to put on a production of *As You Like It,* with (of course) Théodore/Madelaine as the
androgynous Rosalind/Ganymede, d'Albert as the desirous Orlando,
and Rosette as the unrequited Phebe. After much tortured vascillation
and close observation, d'Albert makes up his mind that Théodore is
indeed a woman and declares his love in a letter. Madelaine decides
that he will be an appropriate choice as her first lover, and rewards
his skill in "penetrating" her disguise by making love with him, thus
giving him a single night of "ideal" love. However, Madelaine has unfinished business with the desirous Rosette as well, and spends the
second half of the night of love in *her* room. At daybreak, Madelaine
leaves the château, and the novel ends.

The most obvious and important physiognomical theme in the novel
is of course d'Albert's quest to "read" Théodore's sex, to ascertain the
relation between overt appearance (in this case, clothing) and "true"
identity (in this case, sex and, ultimately, gender). Transvestism, as a
plot device, not only propels the narrative in that it creates a central
riddle to be solved, but more specifically creates the unusual prob-
lem(s) of reading sex and gender, and thereby implicitly puts into
question the very notions of how gender itself is defined.[20] Before
going on to examine these far-reaching implications, however, we
must see how d'Albert's rather limited skills as a physiognomist are
put to the test by the enigmatic figure of Théodore.

Even before d'Albert meets Théodore, the theme of physiognomical analysis is introduced in a scene in which d'Albert is taken by a
friend to a *salon* inhabited by potential mistresses ("une maison, où, à
ce qu'il m'a dit, on voyait un monde de jolies femmes . . . car c'est un
vrai sérail que cette maison-là" [68]). As d'Albert observes the women
in this "harem" and tries to determine everything from their age to
their willingness to enter into a liaison with him from their appearance, we see that his skills as a physiognomist are quite limited. He
cannot seem to make their bodies signify, in spite of the fact that his
friend C*** is quite capable of doing so. As d'Albert himself puts it:
"Mais, mon cher C***, je suis tout à fait neuf sur ces matières-là. Je n'ai
point ce qu'il faut du monde pour distinguer au premier coup d'oeil

une femme frappée avec une qui ne l'est point . . ." (72). C*** counters by pointing out a woman whose physiognomy would be quite revealing, if only d'Albert knew how to read it: "Regarde-moi là-bas un peu, dans ce coin auprès de la cheminée, cette petite femme en rose qui joue avec son éventail: elle te lorgne depuis un quart d'heure avec une assiduité et une fixité tout à fait significatives . . ." (72). Further proving his point that all bodies are legible, C*** reads a woman who has, with great skill and subtlety, been reading d'Albert himself:

> Cette virginale créature, ou je me trompe fort, a déjà supputé dans sa tête ce que les promesses de ta pâleur et de tes yeux noirs pouvaient tenir d'amour et de passion; et ce qui me fait dire cela, c'est qu'elle n'a pas regardé une seule fois de ton côté, du moins en apparence . . . (74–75)

The woman in question is the most debauched of all in the *salon*, but paradoxically also the one with the most innocent appearance. C*** describes her as "un corps de vierge et une âme de fille de joie" (76), a formula which clearly demonstrates the necessity to read, and knowingly interpret, appearances. It is also a formula which prefigures one of the most central concerns of the novel, that of the direct opposition of the physical and the metaphysical domains, a concern which is echoed in d'Albert's self-analysis: "C'est la soirée de ma vie où j'ai eu le plus l'air vertueux et où je l'ai été le moins" (78). This opposition will be even more explicitly thematized and problematized as the narrative progresses.

What interests me most in the *salon* scene, however, is the fact that in this "harem," in this environment of woman-as-object, d'Albert proves himself to be a singularly inept reader of women. This ineptitude, moreover, does not diminish with the experience he acquires as the novel continues, as is clearly demonstrated by his consistent inability to read "la dame en rose" (who becomes "Rosette," his mistress). Furthermore, the notion of the harem, the domain in which women are objects to be scrutinized, manipulated, and selected, is subverted by the introduction of the woman who herself reads d'Albert. The man-as-observing-subject/woman-as-observed-object opposition is clearly exposed as false. Here again, the scene prefigures in a very suggestive way the central concerns of the larger narrative, as we shall see in the subversive figure of Madelaine de Maupin, a woman who is both subject and object of the physiognomical enterprise.

Significantly, d'Albert offers a justification for his physiognomical illiteracy by explaining that his obsessive inner vision of ideal beauty clouds his ability to perceive corporeal reality:

[Cette] tension acharnée de l'oeil de mon âme vers un objet invisible m'a faussé la vue. Je ne sais pas voir ce qui est, à force d'avoir regardé ce qui n'est pas, et mon oeil si subtil pour l'idéal est tout à fait myope dans la réalité;—ainsi j'ai connu des femmes que tout le monde assure être ravissantes, et qui ne me paraissent rien moins que cela. (81)

It is indeed true that d'Albert's "ideal" distorts his perception of reality. However, it is important to give particular attention to the fact that this conflict is not in fact between the physical and the metaphysical realms, but rather between a physical reality and an imagined physicality, between bodies which exist in the material world and a body whose existence is limited to the confines of d'Albert's fantasies. The very use of the term "ideal" is perhaps misleading, as one might infer that metaphysical qualities were the focus of his desire.[21] D'Albert is totally unaware of the oxymoron inherent in the very concept of "ideal beauty," a notion which glibly conflates the material and the immaterial: an ideal *cannot* be incarnated and remain an ideal. On the contrary, however, d'Albert's "ideal" is defined by exclusively physical, esthetic qualities:

Je me suis figuré bien souvent l'endroit qu'elle habite, le costume qu'elle porte, les yeux et les cheveux qu'elle a . . . Elle est d'une taille moyenne . . . elle est plutôt grasse que maigre . . . Elle est blonde avec des yeux noirs,[22] blanche comme une blonde, colorée comme une brune, quelque chose de rouge et de scintillant dans le sourire. La lèvre inférieure un peu large, la prunelle nageant dans un flot d'humide radical, la gorge ronde et petite, et en arrêt, les poignets minces, les mains longues et potelées, la démarche onduleuse . . . les hanches étoffées et mouvantes, l'épaule large, le derrière du cou couvert de duvet . . . elle porte une robe de velours écarlate ou noir avec des crevés de satin blanc ou de toile d'argent, un corsage ouvert, une grande fraise à la Médicis, un chapeau de feutre capricieusement rompu comme celui d'Héléna Systerman, et de longues plumes blanches frisées et crespelées, une chaîne d'or ou une rivière de diamants au cou, et quantité de grosses bagues de différents émaux à tous les doigts des mains. (54)[23]

By his own description, d'Albert has a veritable cult of physical beauty, and he specifies that he admires beauty for its own sake, and not for any power it may have as a signifier of morality:

J'adore sur toutes choses la beauté de la forme;—la beauté pour moi, c'est la Divinité visible, c'est le bonheur palpable, c'est le ciel descendu sur la terre . . .
. . . ô beauté! . . . qui ne pourrait pas s'agenouiller devant toi, pure personnification de la pensée de Dieu!

Je ne demande que la beauté, il est vrai; mais il me la faut si parfaite
que je ne la rencontrerai probablement jamais . . . (133–35)

In other words, d'Albert defines beauty as a virtue in and of itself, and
not as representative of nonphysical virtues. Unlike more traditional
physiognomical notions (like Lavater's), therefore, the body functions
not as a signifier but rather as a self-sufficient, self-referential sign.
Beauty does not *represent* "la pensée de Dieu"; it is a *personnification* of
it: it is itself divinity, in human form.[24]

It is of course fundamental to remember that this is a definition not
of the human body in general, but of the *beautiful* body. For d'Albert,
the body is not necessarily divine—beauty is. It is only at a certain
level of esthetic brilliance (determined of course by d'Albert himself)
that a body comes to be virtuous. It is precisely this phenomenon
which, d'Albert explains, clouds his ability to read other, less than
ideally beautiful, bodies. He cannot interpret them because his own
conception of the human body does not include the body as signifier
of anything other than its own esthetic merit, and even this capacity
would seem to be limited to a body or bodies which exist only in his
imagination and therefore do not *need* to be interpreted.[25] D'Albert's
ability to comprehend and interpret the reality of bodies is thus at
best severely limited. It is this very inability which will prevent him
from reading Madelaine/Théodore, and will therefore form the central
intrigue of the novel.

It is also significant that, for d'Albert, physical beauty is exclusively female. Questions of sexual desire aside, it would be logical to
expect that a cult of physical beauty would include at least the possibility of esthetic attributes in both sexes. However, as d'Albert tells
us, beauty is a solely feminine quality.[26] Men are intended to observe
and appreciate beauty, but only women to possess it:

O beauté! nous ne sommes créés que pour t'aimer et t'adorer à genoux si
nous t'avons trouvée, pour te chercher éternellement à travers le monde
si ce bonheur ne nous a pas été donné; mais te posséder, mais être
nous-mêmes toi, cela n'est possible qu'aux anges et aux femmes. (188)

In typical fashion, d'Albert contradicts himself in the very next paragraph by recounting the tale of having met a young man who had
"la beauté de [ma] laideur." However, in spite of his admiration for
the young man's appearance, he goes on to explain that "ce drôle,
ce n'est que moi un peu mieux réussi et coulé avec un bronze moins
rebelle . . ." (188–89), as if to denigrate the value of the young man's
beauty by suggesting that there was no essential difference between

the young man and himself. Indeed, even d'Albert's desire to be beautiful is modified into a desire to possess beauty in a woman:

> Il m'a toujours semblé que, pour faire ce que je veux (et je ne sais pas ce que je veux), j'avais besoin d'une très grande et très parfaite beauté . . . (138)

> J'ai aimé . . . les femmes, pour posséder au moins dans quelqu'une la beauté qui me manquait à moi-même. (139)

D'Albert's fundamental premise about "real" beauty and sexual identity, in spite of seeming contradictions, therefore remains intact, with men the judges of esthetic value and women the objects of esthetic evaluation.

Obviously, all of this information about d'Albert's esthetic and perceptual preconceptions becomes most relevant to the narrative when he encounters Théodore.[27] In several significant ways, Madelaine/Théodore contradicts d'Albert's most firmly held notions and subverts his entire esthetic system.

The most blatantly obvious way in which Théodore fails to conform to d'Albert's prejudices is in his/her dress. The person who most exactly conforms to d'Albert's esthetic ideal, as it happens, does not wear "une robe de velours écarlate ou noir," but rather men's clothing. D'Albert is understandably completely baffled by this figure who seems "vraiment parfait" in everything but his (apparent) sex. In order to maintain intact his prejudices about beauty as exclusively feminine, his first reaction is to equate the very fact of Théodore's beauty with femininity, by defining his beauty as "feminine" beauty found in a man, rather than allow for the possibility of "masculine" beauty. As we later discover, in the case of Théodore, this is in fact an accurate interpretation; however, in the context of d'Albert's ignorance of Théodore's true sex, it is inescapably oxymoronic. Théodore, as we see in the following description of "his" physical attributes lovingly detailed by d'Albert, is simply *too* beautiful to be a man:

> ce qui me charme le plus est un jeune cavalier . . . il m'a plu tout d'abord, et je l'ai pris en affection, rien qu'à le voir descendre de son cheval. Il est impossible d'avoir meilleure grâce; il n'est pas très grand, mais il est svelte et bien pris dans sa taille; il a quelque chose de moelleux et d'onduleux dans la démarche et dans les gestes, qui est on ne peut plus agréable; bien des femmes lui envieraient sa main et son pied. Le seul défaut qu'il ait, c'est d'être trop beau et d'avoir les traits trop délicats pour un homme. (144)

In seeming contradiction to earlier statements about his amoral love of beauty for its own sake (implying, one would think, that such appreciation might transcend the boundaries of sex), but in keeping with his strictly gendered definition of beauty, d'Albert expresses his regret of Théodore's sex or, significantly, of his own: "Quel dommage que ce soit un homme, ou quel dommage que je ne sois pas une femme!" (144).

As his acquaintance with Théodore continues, his attraction to him grows and he finds himself more and more troubled by his perception of beauty in a man. In order to refute these feelings which, he believes, are "hors nature" and "monstrueux," he tries to convince himself at various moments in the text that Théodore must be a woman. Indeed, he tells himself (and his correspondent) that such great beauty could not exist in a man:

> Il faut que Théodore soit une femme déguisée; la chose est impossible autrement.—Cette beauté excessive, même pour une femme, n'est pas la beauté d'un homme . . . Est-ce que Dieu mettrait ainsi des franges de soie si longues et si brunes à de sales paupières d'homme? Est-ce qu'il teindrait de ce carmin si vif et si tendre nos vilaines bouches lippues et hérissées de poils? Nos os taillés à coups de serpe et grossièrement emmanchés ne valent point qu'on les emmaillotte d'une chair aussi blanche et aussi délicate; nos crânes bossués ne sont point faits pour être baignés des flots d'une si admirable chevelure. (187)

In spite of these very certain pronouncements, however, d'Albert continues to puzzle over the enigma of the sex of his object of desire, without being wholly assured that certain knowledge of the sex of that object would substantially affect the nature or intensity of his desire. Although we as readers already know the answer to the central question, through Madelaine/Théodore's own narrative, a more fundamental, less plot-oriented question remains for us as well: can an esthetic appreciation of human physical beauty override sex? Furthermore, given an appropriately provocative object of esthetic admiration, are our erotic tendencies as sex-specific as we tend to think? Even with our knowledge that d'Albert's desire is in fact a heterosexual one, we are left with the nagging question of "what if": what if Théodore were just as beautiful, but really a man underneath his/her men's clothing? Although the question is being asked in a "safe" way, given our knowledge of Théodore's sex, the question is that of the possibility of bisexuality as an eventual consequence of a philosophy of beauty for beauty's sake. It is this question, and not the already-solved mystery of Théodore's sex, which provides the real tension in the novel. Furthermore, it is a question to which the text offers only the most enigmatic of answers. Without resorting to facile and uninformed biographical speculation about Gautier himself, it is perhaps not unreasonable to ask if such eventualities might not appear logical, if frightening, to a proponent of an absolute, amoral love of physical beauty in all forms. The theme is indeed even more explicitly and suggestively treated in the plot line concerning Madelaine/Théodore and Rosette, as I shall discuss below.

If d'Albert remains somewhat uncertain about Théodore's sex until the very moment in which he/she undresses in front of him, it is interesting to note that he comes closest to certainty when Théodore "disguises" him/herself as Rosalind, for their production of *As You Like It*. The simple definition of clothes as that which conceals is completely reversed as clothes, in this case, *reveal* sexual identity. It is only when dressed as a woman, in her "real" costume, that Madelaine/Théodore can be positively identified as a woman:

> L'image qui jusqu'alors ne s'était dessinée que faiblement et avec des contours vagues, le fantôme adoré et vainement poursuivi était là, devant mes yeux, vivant, palpable, non plus dans le demi-jour et la vapeur, mais inondé de flots d'une blanche lumière; non pas sous un vain déguisement, mais sous son costume réel; non plus avec la forme dérisoire d'un jeune homme, mais avec les traits de la plus charmante femme. (256)

This moment of epiphany, of union between *être* and *paraître*, is, however, curiously destabilized as ambiguity seems to set in once again only a few paragraphs later. Unlike d'Albert, the assembled company does not recognize the truth of the "disguise," and their wonder at Théodore's femininity is expressed through a subtle use of free indirect discourse. The use of the third-person feminine singular pronoun ("elle") by d'Albert to refer to Théodore gives way to a reversion to the masculine pronoun ("il"), and what were described by d'Albert as natural, feminine mannerisms become wondrous in the man that "he" remains to the other observers:

> Elle s'avança dans la chambre, la joue légèrement allumée d'un rouge qui n'était pas du fard, et chacun de s'extasier, et de se récrier, et de se demander s'il était bien possible que ce fût lui, Théodore de Serannes, le hardi écuyer, le damné duelliste, le chasseur déterminé, et s'il était parfaitement sûr qu'il ne fût pas sa soeur jumelle. Mais on dirait qu'il n'a jamais porté d'autre costume de sa vie! Il n'est pas gêné le moins du monde dans ses mouvements . . . C'est prodigieux! C'est inconcevable! L'illusion est aussi complète que possible: on dirait presque qu'il a de la

gorge, tant sa poitrine est grasse et bien remplie, et puis pas un seul poil
de barbe, mais pas un; et sa voix qui est douce! Oh! la belle Rosalinde! et
qui ne voudrait être son Orlando? (257–58)

Indeed, this reintroduction of the possibility of Théodore's being
a man puts into question d'Albert's certainty about the "truth" of
the feminine costume and foreshadows a considerable weakening of
that certainty. When Théodore/Rosalind reappears in men's clothing
(for the *travesti* required by the text of the play), d'Albert's doubts
resurface:

Cela me fit une impression désagréable:—je m'étais si bien accoutumé
déjà à ce costume de femme qui permettait à mes désirs quelques es-
pérances, et qui m'entretenait dans une erreur perfide, mais séduisante!
On s'habitue bien vite à regarder ses souhaits comme des réalités sur la
foi des plus fugitives apparences, et je devins tout sombre quand Théodore reparut sous son costume d'homme, plus sombre que je ne l'étais
auparavant . . . (260)

Upon closer inspection, however, d'Albert changes his reading yet
again, as he perceives that the masculine costume hides a feminine
reality:

Cependant il [son habit] était ajusté de manière à faire pressentir que
ces habits virils avaient une doublure féminine; quelque chose de plus
large dans les hanches et de plus rempli à la poitrine, je ne sais quoi
d'ondoyant que les étoffes ne présentent pas sur le corps d'un homme
ne laissaient que de faibles doutes sur le sexe du personnage . . .
   La sérénité me revint un peu, et je me persuadai de nouveau que
c'était bien effectivement une femme. (260–61)

It is curious, and perhaps in keeping with his acknowledgment of
indulgence in wishful thinking cited above, that d'Albert so readily
accepts this appearance of underlying femininity in the context of
theatrical make-believe. If all theatrical costumes are to some degree
disguises, is it not possible or even likely that Théodore, disguised as
Rosalind disguised as Ganymede, purposely created the appearance
of a woman disguised as a man? Is d'Albert perceiving the reality
of Théodore or merely allowing himself to be deceived by the mind-
boggling complexity of the *mise-en-abîme* of the play being produced?
Perhaps the most thoughtful response to this question is the one
suggested by Ross Chambers' unusual and insightful reading of the
significance of the play within the novel:

il importe que la robe de Rosalinde révèle la féminité de Madelaine, mais il importe davantage que le *rôle* de Rosalinde révèle le caractère indissolublement ambivalent et double, l' "être" synthétique de Madelaine-Théodore . . . Mais si d'Albert reste finalement fort perplexe, c'est que le rôle de Rosalinde est surtout joué en travesti.[28]

Chambers points out in effect that d'Albert's futile attempts to determine visually Théodore's true sex through the costumes of the play are both mirrored and problematized by the thematics of the play itself. The irony of finding out the truth of sexual identity through a play which revolves around sexual ambiguity suggests perhaps that sex is not as simple to define as d'Albert would like to believe, is indeed not necessarily a mere question of anatomy, and that his quest to define it will inevitably become hopelessly embroiled in a play of mirrors.[29]

Furthermore, it is important to note that d'Albert's definition of sexual identity is absolute in its terms: it is a definition which implies a unity of sex and gender, of the physical and the metaphysical, a coherence of mind, body, and soul: a woman has the mind, soul, and body of a woman (as understood by d'Albert himself). The notion of this harmony is particularly ironic coming from d'Albert, a character in whom the opposition and disharmony of mind, body, and soul have been explicitly treated from the beginning of the novel:

> mon âme est pour mon corps une soeur ennemie, et le malheureux couple, comme tout couple possible, légal ou illégal, vit dans un état de guerre perpétuel. (93)

Indeed, d'Albert characterizes his entire cult of beauty as indicative of the opposition between physical and metaphysical concerns:

> mon corps rebelle ne veut point reconnaître la suprématie de l'âme . . . je pense que la correction de la forme est la vertu. La spiritualité n'est pas mon fait, j'aime mieux une statue qu'un fantôme . . . (190)

In spite of these statements, d'Albert conceives with great difficulty the idea that while Théodore's anatomical sex is certainly either male or female, his/her metaphysical self may not correspond to a strictly defined categorization of gender.[30] He denies Théodore's profound androgyny by ascribing gender to his/her mind as well:

> Je me suis épris d'une beauté en pourpoint et en bottes . . . qui vous laisse par moments flotter dans les plus inquiétantes perplexités;—ses traits et

son corps sont celui d'une femme, mais son esprit est incontestablement celui d'un homme. (266)

If Théodore's physical sex can be positively ascertained by the disrobing of his/her genitals, thus effacing his/her androgyny and answering d'Albert's question, a less literal definition, that of gender (which transcends purely anatomical classification), must be left open-ended. Théodore/Madelaine's androgyny is a more profound and suggestive question than mere anatomy can answer. It is precisely this profound ambiguity about the relation between sex and gender that d'Albert so desperately tries to deny throughout the narrative. Madelaine/Théodore's identity cannot merely be defined as the exterior of a man (clothing) covering the interior of a woman (her body), nor even (as d'Albert would have it) the exterior of a woman (her body) covering the interior of a man (her mind). As Kari Weil explains, the distinction between interior and exterior becomes ambiguous as appearance becomes, in some sense, reality:

> Théodore/Maupin . . . is not so easily undressed and appropriated by his/her observer/reader, nor by his/her lover. Neither dress nor language functions for him/her according to the bourgeois code of transparency or superfluity. If the adoption of breeches and a sword in lieu of skirts and petticoats is first thought to be mask or disguise, in a short time that mask becomes the defining element . . . [31]

For her part, Madelaine begins her journey as a "simple" transvestite: a woman in men's clothing. As her travels continue, however, she indeed finds that the distinction between the gender represented by the clothing she has adopted and her "true" gender becomes less and less clear. She herself describes her transformation as that of becoming a man:

> Beaucoup d'hommes sont plus femmes que moi.—Je n'ai guère d'une femme que la gorge, quelque lignes plus rondes, et des mains plus délicates; la jupe est sur mes hanches et non dans mon esprit. Il arrive souvent que le sexe de l'âme ne soit point pareil à celui du corps, et c'est une contradiction qui ne peut manquer de produire beaucoup de désordre . . . Sous mon front poli et mes cheveux de soie remuent de fortes et viriles pensées . . . (288)

At least for the late-twentieth-century reader familiar with the tenets of contemporary feminist thought, the clearest implication of Madelaine's self-definition is that it indirectly (and, perhaps, inadvertently) subverts reductive notions of sex and gender. Because Madelaine has come to react to the world and thus to see the world in ways which are traditionally defined as masculine, she assumes that such ways are masculine traits within a woman, not recognizing the fundamental oxymoron implied by that notion. If she, a woman, experiences such thoughts, they are necessarily not "masculine." The mind housed in a woman's body is by definition "feminine," regardless of its relation to the traditional societal constructs of gender definition. Madelaine/Théodore is truly androgynous, an individual in whom the coexistence of "masculine" and "feminine" qualities puts into question the very possibility of such categories, as Kari Weil has so suggestively stated:

> in his/her transgression of gender boundaries, s/he defies categorical thinking based on a return to identity of the same. S/he is not either/or, but both/and . . . Resisting the bounds of identity based on prior models of male and female, Madelaine/Théodore reveals the illusory nature of their "identity" as polar opposites.[32]

If the character of Madelaine/Théodore subverts definitions of the masculine and the feminine, he/she also subverts the opposition between (male) subject and (female) object. As we saw above, d'Albert's esthetic notions clearly divide the esthetic enterprise along gender lines: men are observing, evaluating, loving subjects, while women are observed, evaluated, beloved objects. It is precisely in order to escape the ignorant and disempowered state forced on women by such a system that Madelaine first dresses as a man and sets out on her journey.

Madelaine's quest is a quest for specifically that knowledge which has been denied her as a woman, knowledge of men. She senses moreover that men must be read physiognomically, but lacks the knowledge with which to engage in such reading:

> je pressentais dans leur vie beaucoup de côtés défectueux et obscurs, soigneusement voilés à nos regards, et qu'il nous importait beaucoup de connaître; quelquefois, cachée derrière un rideau, j'épiais de loin les cavaliers qui venaient à la maison, et il me semblait alors démêler dans leur allure quelque chose d'ignoble et de cynique, une insouciance grossière ou une préoccupation farouche que je ne leur retrouvais plus dès qu'ils étaient entrés, et qu'ils semblaient dépouiller comme par enchantement sur le seuil de la chambre. Tous, les jeunes comme les vieux, me paraissaient avoir adopté uniformément un masque de convention, des sentiments de convention et un parler de convention lorsqu'ils étaient devant les femmes. (207–8)

Madelaine is perceptive enough to recognize the existence of a sign system which she cannot understand. She is also enough of a physiognomist to have perceived this fact through careful scrutiny of men's faces:

> je suivais de l'oeil . . . les groupes bourdonnants et rieurs de jeunes gens qui . . . reprenaient leur promenade tout en causant et en jetant au passage des oeillades ambiguës. Sur leurs bouches dédaigneusement bouffies voltigeaient des ricanements incrédules; ils avaient l'air de se moquer de ce qu'ils venaient de dire, et de rétracter les compliments et les adorations dont ils nous avaient comblées. Je n'entendais pas leurs paroles; mais je comprenais, au mouvement de leurs lèvres, qu'ils prononçaient des mots d'une langue qui m'était inconnue et dont personne ne s'était servi devant moi. Ceux mêmes qui avaient l'air le plus humble et le plus soumis redressaient la tête avec une nuance très sensible de révolte et d'ennui . . . (208)

Unlike a traditional (Lavaterian) physiognomist, who uses the language of the body as a means to knowledge of the metaphysical nature of man, Madelaine uses facial expression as the signifier of a signified which is itself a language. By studying men's faces, she realizes that they speak another language among themselves than when in the presence of women. Madelaine's knowledge of this language would in turn allow her to interpret their facial expressions as signifiers of the all-male code. In other words, she must have access to male verbal expression in order to decipher male facial expression.

It is particularly necessary for Madelaine to understand the male language, because she knows that she, as a woman, provides the content of their discourse:

> Souvent je comprenais, à de certaines attitudes, à quelques gestes détournés, à des coups d'oeil lancés obliquement, qu'il était question de moi et que l'on parlait de mon âge ou de ma figure. (209)

Because Madelaine wants to know men in order to choose the right one, she must reverse the seemingly irreversible roles of man-as-subject, woman-as-object. She brings about this feat of course by pretending to be a man, thus becoming privy to the masculine discourse about women. Madelaine must become in her turn an observing, evaluating subject. She defines her agenda in pseudoscientific, Balzacian terms:

> je voulais étudier l'homme à fond, l'anatomiser fibre par fibre avec un scalpel inexorable et le tenir tout vif et tout palpitant sur ma table de

dissection; pour cela il fallait le voir seul à seul chez lui, en déshabillé, le suivre à la promenade, à la taverne et ailleurs. (210)

Furthermore, Madelaine is well aware of the reasons why she must "become" a man in order to be an observant, knowledgeable subject. Women, she tells us, are systematically forbidden access to knowledge and denied status as thinking subjects in her world:

> C'est une chose effrayante à penser et à laquelle on ne pense pas, com- bien nous ignorons profondément la vie et la conduite de ceux qui pa- raissent nous aimer et que nous épouserons . . .
> Il nous est défendu de prendre la parole, de nous mêler à la conver- sation autrement que pour répondre oui ou non, si l'on nous interroge. Aussitôt que l'on veut dire quelque chose d'intéressant, l'on nous renvoie étudier notre harpe ou notre clavecin . . .
> A force de vouloir nous empêcher d'être romanesques, l'on nous rend idiotes. Le temps de notre éducation se passe non pas à nous apprendre quelque chose, mais à nous empêcher d'apprendre quelque chose.
> Nous sommes réellement prisonnières de corps et d'esprit . . . (210– 12)[33]

Madelaine also explains that it is this system of repression that not only renders women incapable of being thinking subjects, but also renders them such facile objects of observation. It is because they are denied the power to decide, choose, or even think that they are so very easily read by men:

> Nous autres, notre vie est claire et se peut pénétrer d'un regard.—Il est facile de nous suivre de la maison au pensionnat, du pensionnat à la maison;—ce que nous faisons n'est un mystère pour personne . . .
> Le cristal le plus limpide n'a pas la transparence d'une pareille vie.
> . . . Notre vie n'est pas une vie, c'est une espèce de végétation comme celle de la mousse et des fleurs . . . (211)

What is remarkable in Madelaine, of course, is her ability to ques- tion the rigid system that has denied her knowledge in general and more specifically that of the opposite sex. She is indeed a "prisonnière de corps," but clearly not a "prisonnière d'esprit." Madelaine has the clear-sightedness to recognize that the subject/object distinction be- tween men and women does not reflect an essential difference, but is merely a social construct. She realizes that she, given the chance, would be just as capable of reading men as they have been of read- ing her. The irony, of course, is that she cannot accomplish this as a woman and must become a man, at least in appearance, in order to become a subject, rather than an object, of physiognomical interpretation. Madelaine accepts this irony, however, as she is less concerned with changing the role of women in her society (indeed, it would be hopelessly anachronistic of us to ascribe such an agenda to her) than with her personal acquisition of knowledge. She must adopt a mask in order to see beyond others' masks. As Anne Bouchard puts it:

> son propre masque est d'abord une nécessité tactique. Si elle choisit de s'avancer masquée, sur le terrain des masques, c'est qu'à l'abri de son pseudonyme et de son déguisement, elle pourra déjouer plus facilement les feintes, et mener plus à fond son enquête, le masque passant alors, dans un renversement dialectique, au service de la vérité . . .[34]

Madelaine recognizes, furthermore, that her new life and new clothes are more than a mere mask and symbolize a profound and irreversible transition:[35]

> avec mes robes et mes jupes, j'avais laissé mon titre de femme; dans la chambre où j'avais fait ma toilette étaient serrées vingt années de ma vie qui ne devaient plus compter et qui ne me regardaient plus. Sur la porte on eût pu écrire:—Ci-gît Madelaine de Maupin; car en effet je n'étais plus Madelaine de Maupin, mais bien Théodore de Sérannes . . .
> Le tiroir où étaient renfermées mes robes, désormais inutiles, me parut comme le cerceuil de mes blanches illusions;—j'étais un homme, ou du moins j'en avais l'apparence: la jeune fille était morte. (213)

It is indeed true that the notion of a mask, a disguise completely distinct from its wearer's true identity, is a figure which becomes increasingly inappropriate to describe Madelaine's adoption of men's clothing. As the narrative continues, she becomes more and more at ease with the conventionally masculine poses and habits she must take on to render her identity as Théodore plausible. Not only do her riding, shooting, and swordsmanship improve remarkably with so much practice, but more unsettling and significant aptitudes develop as well, as we shall see below.

Given the freedom accorded by her new status as a man, Madelaine fully exercises her ability to read and judge others. She demonstrates from her very first encounter as Théodore with a group of men that she is already secure in her ability to read at least some basic physiognomical signs:

> Plusieurs cavaliers entrèrent dans l'auberge . . .
> —Ils étaient tous jeunes, et le plus âgé n'avait assurément pas plus de

trente ans: leurs vêtements annonçaient qu'ils appartenaient à la classe supérieure, et, à défaut de leurs vêtements, la facilité insolente de leurs manières l'eût fait assez comprendre. Il y en avait un ou deux qui avaient des figures intéressantes; les autres avaient tous, à un degré plus ou moins fort, cette espèce de jovialité brutale et d'insouciante bonhomie que les hommes ont entre eux, et dont ils se dépouillent complètement lorsqu'ils sont en notre présence. (219–20)

Her knowledge of the true identity of men becomes much more profound as the evening progresses, and she finally becomes privy to the all-male discourse about women she has always longed to hear. She finds to her chagrin that this masculine language is crude, disrespectful, and deeply misogynist. She discovers that men do indeed objectify women, and that consequently the subject/object distinction between the sexes that necessitated her journey in the first place is more firmly entrenched than perhaps even she had thought. She discovers specifically that men are much more concerned with the beauty of a woman's body than with that of her mind or soul, an attitude to which the discourse of one of her tablemates in the *auberge* clearly attests:

—Elle est folle de moi:—c'est la plus belle âme du monde; en fait d'âmes, je m'y connais; je m'y connais aussi bien qu'en chevaux pour le moins, et je vous garantis que celle-là est une âme première qualité . . . mais elle n'a presque pas de gorge, elle n'en a même pas du tout, comme une petite fille de quinze ans au plus.—Elle est assez jolie du reste; sa main est fine, et son pied petit; elle a trop d'esprit, et pas assez de chair, et il me prend des envies de la planter là. Que diable on ne couche pas avec des esprits. (223)

Lest there be any doubt that she knows what such a discourse implies about men's attitudes, Madelaine lets her reader know that she recognizes the significance of what she has learned:

La conversation dura encore quelque temps, la plus folle et la plus dévergondée du monde; mais, à travers toutes les exagérations bouffonnes, les plaisanteries souvent ordurières, perçait un sentiment vrai et profond de parfait mépris pour la femme, et j'en appris plus dans cette soirée qu'en lisant vingt charretées de moralistes. (223–24)

The danger of the game that Madelaine plays, however, is that her knowledge quickly transforms her into a sort of oxymoronic being. Just as her transvestism places her in a gray area between the two sexes, so her knowledge of the reality of men makes her different from other women. If women in her society are defined by their ignorance and status as objects, what is a woman who possesses precisely that knowledge that is systematically denied to all women?

Furthermore, how can she reconcile her growing repugnance for men's attitudes with an equally thriving attraction to their bodies? Madelaine is both cognizant of, and articulate about, this contradiction, as she tells her reader when she is obliged to share a bed with one of her (male) dinner companions:

> Assurément ce que j'avais entendu n'était pas de nature à me prédisposer à la tendresse et à la volupté:—j'avais les hommes en horreur.—Cependant j'étais plus inquiète et plus agitée que je n'aurais dû l'être: mon corps ne partageait pas la répugnance de mon esprit autant qu'il l'aurait fallu. (226–27)

Not unlike d'Albert, Madelaine is caught in a web of irreconcilable contradiction between her mind and her body. Ironically, she has already, as a "man," internalized misogynist notions about the unbridled corporality of women, and thus laments her feminine "weakness" for sexuality:

> Une effervescence subite, un bouillon de sang peut-il à ce point mater les résolutions les plus superbes? et la voix du corps parle-t-elle plus haut que la voix de l'esprit? . . . Je commence à être de l'avis des hommes: quelle pauvre chose que la vertu des femmes! et de quoi dépend-elle, mon Dieu! (228)

The most significant way in which Madelaine's "masculine" persona comes to be very real, however, is in her relations with other women. As a man, she is granted rights as a thinking, evaluating, desiring subject; the objects of her "masculine" desire must of course, according to the rules of compulsory heterosexuality, be women. This complex situation is fully enacted in her relation with Rosette, sister of one of her traveling companions. At first, Madelaine/Théodore courts Rosette out of pure *politesse*. However, as Rosette begins to respond more and more forcefully to these advances (and indeed, to make more than a few advances of her own), Madelaine/Théodore must not only find a way to politely preempt any real sexual encounter but also face her own growing desire.

As Rosette's attempts to seduce Madelaine/Théodore become more and more bold, so Madelaine/Théodore's feelings become more and more ambiguous. In two titillating scenes between the two women, the reader (like the women) is teased and excited by the mounting purposefulness of the sex play, only to have the scene interrupted at its peak by Rosette's brother, who makes a habit of bursting into rooms at precisely the wrong (right?) moment. The reader is thus left with a sense of frustration and curiosity which mirrors Madelaine/ Théodore's erotic confusion.

In keeping with her pattern of intelligence and self-awareness, Madelaine/Théodore is quite lucid about her new feelings, even if she is more willing to conceptualize them as esthetic than as erotic. She recognizes that, as a "man," she has come to see women as esthetic objects and to appreciate their beauty in a way she never could have as a woman:

> depuis que j'ai quitté les habits de mon sexe et que je vis avec les jeunes gens, il s'est developpé en moi un sentiment qui m'était inconnu:—le sentiment de la beauté. Les femmes en sont habituellement privées, je ne sais trop pourquoi, car elles sembleraient d'abord plus à même d'en juger que les hommes;—mais, comme ce sont elles qui la possèdent, et que la connaissance de soi-même est la plus difficile de toutes, il n'est pas étonnant qu'elles n'y entendent rien. (296)

Perhaps the most intriguing feature of the passage above is the fact that Madelaine seems to accept, as does d'Albert, the notion that beauty is necessarily and exclusively a feminine quality. She says that women cannot judge beauty, since self-knowledge is the most difficult, without ever allowing for the possibility of masculine beauty. Following her logic, one would assume that, were masculine beauty to exist, only women would be capable of judging and appreciating it. However, such an eventuality is never even hinted at. Her desire for men, in contrast to her feelings for Rosette, is never expressed as a possibly logical consequence of esthetic appreciation, but rather as a vague, dark, physiological drive.

For all her lucidity and desire to reject a system in which men are subjects and women objects, Madelaine nonetheless would seem to concur with the system, at least in the domain of esthetics.

She states her own appreciation of feminine beauty, and dismissal of masculine physical attributes, in no uncertain terms:

> Je la contemplai quelque temps avec une émotion et un plaisir indéfinissable, et cette réflexion me vint que les hommes étaient plus favorisés que nous dans leurs amours, que nous leur donnions à posséder les plus charmants trésors, et qu'ils n'avaient rien de pareil à nous offrir.— Quel plaisir ce doit être de parcourir de ses lèvres cette peau si fine et si polie, et ces contours si bien arrondis, qui semblent aller au-devant

du baiser et le provoquer! ces chairs satinées, ces lignes ondoyantes et qui s'enveloppent les unes dans les autres, cette chevelure soyeuse et si douce à toucher; quels motifs inépuisables de délicates voluptés que nous n'avons pas avec les hommes! (296)

Although it is quite clear to the reader of the passage above (at least in the late twentieth century) that what Madelaine/Théodore is approaching is full-blown homosexual desire, she herself is less than sure how to define her new feelings:

je l'avouerai à ma honte, cette scène [avec Rosette], tout équivoque que le caractère en fût pour moi, ne manquait pas d'un certain charme qui me retenait plus qu'il n'eût fallu; cet ardent désir m'échauffait de sa flamme, et j'étais réellement fâchée de ne le pouvoir satisfaire: je souhaitai même d'être un homme, comme effectivement je le paraissais, afin de couronner cet amour, et je regrettai fort que Rosette se trompât. (299)

Madelaine will later discover, however, what we already know: that she doesn't need to be a man in order to "crown that love." In other words, she will discover that physical expressions of homosexual desires are not in fact "impossible" (as d'Albert would have it), and that sexual desire for another woman does not put her "true" gender into question. At the time of her first interrupted love scene with Rosette, however, Madelaine literally cannot envisage what might have taken place between the two of them:

un attrait invincible me fit revenir en avant, et je le lui rendis presque aussi ardent qu'elle me l'avait donné. Je ne sais pas trop ce que tout cela fût devenu . . . (300)

un quart d'heure plus tard, le diable m'emporte si je sais le dénouement qu'aurait pu avoir cette aventure,—je n'y vois pas de possible . . . il aurait bien fallu que cela finît d'une manière ou de l'autre. (301–2)

All that is clear to Madelaine is that Rosette is languishing for her (or, to be more precise, for Théodore), and that she herself has experienced, and continues to experience, a "sensation étrange" and a "désordre inconcevable" when near Rosette.

In spite of her inability to define her own feelings, Madelaine has no trouble whatsoever reading Rosette's unrequited love and desire from abundant physiognomical clues:

Une pâleur subite couvrit sa belle figure: elle me jeta un regard douleureux et plein de reproches . . . (302)

Les couleurs que l'annonce de mon départ avaient chassées de ses joues n'y revinrent pas aussi vives qu'auparavant;—il lui resta de la pâleur sur la joue et de l'inquiétude au fond de l'âme. (303)

elle en était visiblement affectée: une expression de tristesse inquiète avait remplacé le sourire toujours frais épanoui de ses lèvres; les coins de sa bouche, si joyeusement arqués, s'étaient abaissés sensiblement, et formaient une ligne ferme et sérieuse; quelque petites veines se dessinaient d'une manière plus marquée à ses paupières attendries; ses joues, naguère si semblables à la pêche, n'en avaient conservé que l'imperceptible velouté. (305)

sa paleur augmentait chaque jour . . . (306)

Le souvenir des temps qui n'étaient plus et qu'elle regrettait donnait à sa figure une mélancolique expression d'attendrissement. (307)

Aside from the remarkable density of physiognomical commentary (it is hard to imagine five pages of a novel by anyone other than Balzac containing as many references to a single character's physiognomy), there are several notable aspects of Madelaine's physiognomical reading of Rosette. The first of these is the fact that, because of her having adopted the persona of a man and in contrast to the original intent of her "aventure," Madelaine finds herself in a situation in which she must read a woman rather than a man. With the distance her masculine clothes give her from members of her own sex, Madelaine is put in the position of interpreting and evaluating other women; because this is traditionally a man's role, she begins to formulate a self-definition which emphasizes her difference from other women.

When referring to the various personae Rosette adopts in her attempts at seduction, Madelaine speaks of "tous ces adorables masques qui vont si bien aux femmes, qu'on ne sait plus si ce sont de véritables masques ou leurs figures réelles" (304). Perhaps without even being fully aware of it, Madelaine has ceased to use the first person plural ("nous," "nous autres") when speaking about women, replacing it with the distanced third person plural ("elles"). She clearly identifies with the "on" of "on ne sait plus," that is to say those who observe and attempt to interpret women—men. The masks ("véritables" or other) are ostensibly adopted in order to confuse and entice prospective lovers, who are presumed to be male. The difference in Madelaine's situation is of course that she is in the oxymoronic position of being a "male" observer/lover without being a male. As we saw above, her status as both observer and lover of women is not entirely a ruse: Madelaine does indeed both interpret and desire a woman.

Not yet understanding, as she will eventually, that it is in fact possible to be a woman who desires another woman, she believes that her own true gender identity is being effaced ("je l'aimais réellement beaucoup, et plus qu'une femme n'aime une femme" [310]). She believes that she is, in some essential sense, becoming a man. Indeed, like d'Albert, she even wishes to be the opposite sex in order to satisfy the seemingly insoluble predicament of homosexual desire:

> souvent, assise auprès de Rosette, sa main dans ma main, l'entendant me parler avec son doux roucoulement, je m'imagine que je suis un homme, comme elle le croit . . . (307)

Whereas d'Albert's desire for Théodore functions as a "safe" way to introduce the question of homosexual desire in the text (as soon as the reader knows Théodore's true identity as Madelaine), the relation between Madelaine and Rosette is much more suggestive. Madelaine's desire for Rosette begins to seem quite genuine; furthermore, it is not at all certain that knowledge of Madelaine's true sex would lessen Rosette's desire.[36]

It is precisely at this point, the moment of the most heightened tension about sexuality and gender identity the reader has yet experienced in the text, that Gautier teases him/her by breaking off Madelaine's narrative and returning to the less compelling d'Albert. The chapters are structured in such a way as to prolong the narrative tension, and thus underscore the importance of what I would argue are the most compelling questions of the novel, the questions to which all others in the text logically lead: what is the relation of esthetics, erotics, and ethics? and what is the relation between sexual desire and gender identity?

True to his commitment to ambiguity, Gautier provides us not with answers to these questions but rather with suggestions. On the one hand, the suspense (or, one might say, titillation) surrounding the disrobing of Madelaine/Théodore is laid to rest with the conclusion of the novel, the sexual fulfillment of d'Albert's desire for him/her. At last d'Albert can actually see the body which has provoked his desire, define, once and for all, his desire as "natural" (heterosexual), and thus set the world straight (no pun intended) and free of ambiguity.[37] The very sight of Madelaine's nude, female body realigns sexual and gender identity for d'Albert: because the person he has desired is in fact all woman, he is therefore all man. Furthermore his ability to recognize a real woman, even under men's clothing, seems to reaffirm his own identity as a masculine subject of physiognomical interpretation. Indeed, Madelaine herself characterizes her decision to go to bed with d'Albert as the ultimate reward for his assiduity and physiognomical skill:

> il se mit à m'observer et à m'étudier avec l'attention la plus minutieuse; il doit connaître particulièrement chacun de mes cheveux et savoir au juste combien j'ai de cils aux paupières; mes pieds, mes mains, mon cou, mes joues, le moindre duvet au coin de ma lèvre, il a tout examiné, tout comparé, tout analysé, et de cette investigation où l'artiste aidait l'amant il est ressorti, clair comme le jour (quand il est clair), que j'étais bien et dûment une femme, et de plus son idéal, le type de sa beauté, la réalité de son rêve . . . (355)

> Puisque d'Albert m'a reconnue sous mon travestissement, il est bien juste qu'il soit récompensé de sa pénétration; il est le premier qui ait deviné que j'étais une femme, et je lui prouverai de mon mieux que ses soupçons étaient fondés. (357)

We should remind ourselves, however, that Madelaine's reading of d'Albert is in fact much more exact and confident than his of her. Until his very viewing of her nude body, d'Albert was not *certain* of his ability to read sexual identity. A certain nagging measure of ambiguity about Théodore's sex, and consequently about his own sexual identity as well, stays with d'Albert until the final pages of the novel.

What makes *Mademoiselle de Maupin* the odd and fascinating text it is, however, is precisely the lack of resolution, the undeniable uncertainty which lingers even after the consummation of d'Albert's desire. The unveiling of Madelaine's body and the satisfaction of d'Albert's desires to know and possess Madelaine/Théodore, seemingly fitting conclusions to the problems posed by the plot, are in fact false conclusions.[38] The real conclusion of the novel comes *after* this scene of illumination and supposed resolution, when Madelaine leaves d'Albert's room and enters Rosette's.

We as readers, as is the case from the very introduction of Madelaine as narrator, are in possession of knowledge greater than d'Albert's. If d'Albert has discovered Madelaine's "true" sex in the most literal, anatomical sense, he has not yet discovered what "true" gender may be, and what the relation between (physical) sex and (metaphysical) gender may imply. Throughout the novel, homosexual desire has been equated with gender confusion: if d'Albert loves Théodore, he must be (in some essential but ill-defined sense) a woman; if Madelaine desires Rosette, she must be a "man." The equation of sexual identity and an essentialist definition of gender would lead to the logical conclusion that Madelaine, through both her anatomy and her having made love to d'Albert, is a "real" woman. Her crossing over (both literally and figuratively) into Rosette's room puts d'Albert's newly (re)found certainty about gender into question: if Madelaine has the body of a "real" woman, and the desires of a "real" woman (in other words, desire for a man), what are we to make of the consummation of her desire for Rosette?

Although Madelaine defines herself as belonging to a "troisième sexe à part," we as twentieth-century readers know to read this as merely an antiquated code word for homo- or bisexuality. Is it not, ultimately, Madelaine rather than d'Albert who carries a love of beauty for beauty's sake to its logical, gender-blind conclusion by loving both of the beautiful objects (which happen to be human bodies of different sexes) in her path?

What the surprising nonconclusion of the narrative represents is the ultimate failure of reading on d'Albert's part. Even after the anatomical sex of the object of his desire has been established, even after he has seen with his own eyes her nude body, he interprets what he sees incorrectly by insisting on ascribing more than anatomical meaning to anatomical reality. He knows that Madelaine is a woman, but oversimplifies the relations between sex and gender, and between gender and sexuality,[39] and thus does not know what sex can (and cannot) signify, what gender may (or may not) be. Indeed, neither do we, according to the information provided by the text: we do know more than d'Albert in that we know what gender isn't (it isn't necessarily an absolute criterion in the choice of objects of desire), but the text leaves us with an undeniable ambiguity concerning any real definition of gender, in fact any signification of sexual identity beyond the purely anatomical.

If the novel suggests the possibility of bisexuality, and teases its more voyeuristic readers with a few scenes of frustrated homosexual seduction, it does not enlighten us as to the actual physical reality of such a possibility. The final love scene between Rosette and Madelaine, like those which precede it, lies beyond that which can be narrated, and is thus alluded to in only the most coy fashion by the unidentified third-person narrator:

> Au lieu de rentrer dans sa propre chambre, elle entra chez Rosette.—Ce qu'elle y dit, ce qu'elle y fit, je n'ai jamais pu le savoir, quoique j'aie fait les plus consciencieuses recherches.—Seulement une femme de chambre de Rosette m'apprit cette circonstance singulière: bien que sa maîtresse n'eût pas couché cette nuit-là avec son amant, le lit était rompu et défait, et portait l'empreinte de deux corps.—De plus, elle me montra deux

perles, parfaitement semblables à celles que Théodore portait dans ses
cheveux en jouant le rôle de Rosalinde. (368–69)

The narrator, with an unmistakable wink in his discourse, leaves the
episode open-ended, for us to puzzle over:

> Je livre cette remarque à la sagacité du lecteur, et je le laisse libre d'en
> tirer toutes les inductions qu'il voudra; quant à moi, j'ai fait là-dessus
> mille conjectures, toutes plus déraisonnables les unes que les autres, et
> si saugrenues que je n'ose véritablement les écrire, même dans le style
> le plus honnêtement périphrasé. (369)

Although he dares not narrate his conjectures, the narrator leaves
us with our final and easiest clues to read; although Madelaine's body
has left the château, she has literally left her mark (in the bed) to be
read. The two quasi-corporeal signifiers, the mark of her body in the
bed and the pearls, are fairly clear indicators of the basic scenario.
In spite of a fundamental difference in narratability, what transpired
between Madelaine and Rosette is directly comparable to what transpired between Madelaine and d'Albert; it was nothing less (nor more)
than the consummation of erotic desire.

In *La Comédie au château*, Ross Chambers argues that there is no
such symmetry in the conclusion of the novel:

> L'insistance que Gautier met à rappeler ainsi le caractère double de l'idéal
> alors même qu'on le voit s'offrir à d'Albert sous la forme enfin accessible
> d'une femme, est de la plus grande importance pour qui veut apprécier
> la très forte nuance d'ironie qui marque le dénouement. La perplexité
> du jeune homme sera dissipée complètement par la scène d'amour finale
> où Madelaine laissera tomber enfin toutes ses voiles. Plus de doutes,
> semble-t-il, sur le sexe de "Rosalinde." Mais faut-il rappeler que la dialec-
> tique du livre nous avait amené à la conclusion qu'il faut un être double
> (Rosette-d'Albert) pour aimer cette créature double qu'est Mademoi-
> selle de Maupin? Or, d'Albert-homme finit par aimer Madelaine-femme.
> Mais que devient de son côté Rosette-femme amoureuse de Théodore-
> homme? Malgré une scène d'amours lesbiennes que Gautier suggère
> afin de faire pendant aux amours de Madelaine et de d'Albert, Rosette
> est en fait exclue du bonheur par ce qui fait précisément le bonheur de
> d'Albert.[40]

Although I agree completely with Chambers' recognition of the heavy
dose of irony which colors the dénouement, and with his characterization of the effect of the final love scene on d'Albert's "perplexity,"
I cannot agree with his reading of the significance of the love scene

between the two women. Chambers is, however, absolutely right in reminding us that the dialectic of the novel brings us to the conclusion that Madelaine can be truly satisfied only by "un être double."

Chambers goes on to characterize Rosette's exclusion from happiness as directly analogous to Phebe's in *As You Like It.* I would argue that he has failed to recognize the fundamental disparity between the story of Madelaine and Rosette and that of Rosalind and Phebe: whereas Phebe's desire for Rosalind/Ganymede is truly unrequited, Rosette's desire for Madelaine/Théodore is, ultimately, amply reciprocated. Madelaine is ultimately a sort of anti-Rosalind: while Rosalind repeats the refrain "And I for no woman" in reference to questions about his/her love and desire, Madelaine's responses to Rosette's physical advances become increasingly less ambiguous and more enthusiastic as the novel progresses. Rosemary Lloyd defines *As You Like It* as an "antimodel" for *Mademoiselle de Maupin:*

> *As You Like It* is less a model than an *antimodel* for *Mademoiselle de Maupin:* while the play offers a happy ending for Rosalinde and Orlando . . . and a less happy one for Phoebe, who must make do with Silvius, Madelaine's refusal to limit herself to one sex forces the other main characters to seek consolation in each other, but also of course to perpetuate her as a bisexual being by speaking her name in a kiss.[41]

What Lloyd sees (and Chambers does not) is that Madelaine, always more practical and less blinded by idealized desire than d'Albert, finds a very real solution to her problem. Because Madelaine's emotional and erotic needs must be fulfilled by both a woman and a man, and because of the anatomical improbability of literal hermaphrodism, she finally recognizes, with characteristic lucidity, that the best possible option is that of bisexuality. Having had deeply erotic feelings about Rosette for some time, she ultimately expresses those feelings. After having lost her (heterosexual) virginity with d'Albert and thus acquired at least a certain part of the knowledge she was seeking, she proceeds to apply that knowledge to homosexual lovemaking with Rosette. She has indeed found the closest thing to the mythical "être double" she can.

Although Chambers characterizes the lesbian scene that Gautier "suggests" as merely serving as a "pendant" to d'Albert's scene with Madelaine, I would suggest rather that it is perhaps an even more significant scene: it is the suggestion of the actual possibility, and reality, of homosexual relations that provides the true dénouement of the narrative. The idea that has been toyed with for so long is finally acted out in the *histoire,* if not in the *récit.* If homosexuality remains unnarratable, it does not remain outside the realm of human possibility as represented in the text. Indeed, the very fact of not narrating the lesbian scene serves to underscore its importance. It is as if the entire text has been circling around the question of the possibility or impossibility of homosexuality, in order to end with the open-ended, but nonetheless clearly implied, response suggested by the quasi-corporeal clues left behind by Madelaine. D'Albert is ultimately proved wrong yet again: his repeated characterizations of physical desire for a member of the same sex as "impossible" are definitively subverted by the surprising dénouement.

Kari Weil gives perhaps the most far-reaching and perceptive reading to date of the implications of the conclusion of the novel:

> Leaving d'Albert's room on the night of the consummation of their love to enter the room of his former lover, Rosette, s/he subverts both narrative and social expectations. S/he enters a realm which marks the limits of d'Albert's foresight, the limits of narrative truth, and, more importantly, the limits of representation in its acceptable, masculine form . . .
>
> In that private space of non-representation, Maupin/Théodore thus inscribes both self and sex, not in terms of the either/or which d'Albert and the reader have expected, but outside the binary logic of polarities, as a both/and, "un sexe qui n'a pas encore de nom" (356).[42]

Weil goes on to articulate an analogy between the erotic and esthetic domains, explaining the significant implications of the erotic conclusion in relation to esthetic considerations:

> Protesting the repressive nature of desire, whether aesthetic or erotic, whether classical or romantic, which satisfies only "une seule face de [son] caractère" (356), s/he envisions love as a process of endless self-fragmentation and creation; "Ma chimère serait d'avoir tour à tour les deux sexes pour satisfaire à cette double nature:—homme aujourd'hui, femme demain, . . . Ma nature se produirait ainsi tout entière au jour, et je serais parfaitement heureuse, car le vrai bonheur est de se pouvoir développer librement en tout sens et d'être tout ce qu'on peut être" (357). S/he thus introduces d'Albert and the reader to a new and modernist relationship with art and beauty, and to a new type of critical reading, one which emphasizes a Nietzschean idea of process without totalization or conclusion.[43]

Rather than take Weil's reading one step further into profundity, I would like to take it one step back by emphasizing what is obvious but not explicit in her statement. Madelaine's conception of love is not only "a process of endless self-fragmentation and creation"; it is also, seen in much more practical, physical terms, bisexual. Her protest against the "repressive nature of desire" is not only ontological, but also sociopolitical: it is a protest against the fact that she cannot, according to her culture, consummate her intense feelings of physical attraction to Rosette. As Weil tells us above, Madelaine indeed represents a "both/and" rather than an "either/or." At the risk of belaboring the obvious, I would like to emphasize that this "both/and" refers in part to Madelaine's ultimate expression of her bisexual desires. The conclusion of the novel is indeed the realization of her "chimère," her fantasy of bisexuality, of having as lovers "homme aujourd'hui, femme demain." Madelaine introduces the reader not only to a modernist way of reading, but also to a decidedly modern vision of sexual politics.[44] In this way, as in others, Madelaine gives a lesson to d'Albert. If it is he who produces the pompous rhetoric of an absolute estheticism but shrinks from its logical conclusion of polymorphous eroticism, it is she who ultimately has the courage to deny the boundary between esthetics and erotics, and in so doing, deny conventional morality, as d'Albert so longs to do.

It is important for us to remember here d'Albert's earlier insistence on his absolute and amoral estheticism:

> Peu à peu ce qu'il y avait d'incorporel s'est dégagé et s'est dissipé, il n'est resté de moi qu'une épaisse couche de grossier limon . . . le monde de l'âme a fermé ses portes d'ivoire devant moi: je ne comprends plus que ce que je touche avec les mains . . . Il n'y a plus, hélas! qu'une chose qui palpite en moi, c'est l'horrible désir qui me porte vers Théodore.—Voilà où se réduisent toutes mes notions morales. Ce qui est beau physiquement est bien, tout ce qui est laid est mal. (198–99)

> Comme on ne cherche que la satisfaction de l'oeil, le poli de la forme et la pureté du linéament, on les accepte partout où on les rencontre. C'est ce qui explique les singulières aberrations de l'amour antique. (200)

Of course, even the passages above reveal the limits of d'Albert's "absolute" estheticism and the very real sense of confusion and shame he feels about his attraction for Théodore ("l'horrible désir," "les singulières aberrations de l'amour antique"). However, as the text progresses, it becomes increasingly even clearer that d'Albert's stance is merely a pose, and that for him beauty does not merely refer to itself (as he would have us believe he believes), but rather functions as a signifier, at least of sex. Furthermore, for d'Albert, the sex of one's object of esthetic and erotic desire is clearly a moral question.[45] D'Albert's need to read sexual identity is directly proportional to his desire for the androgynous figure of Théodore. Beauty, it would seem, cannot merely signify itself without "moral" implications.[46] The question, however, remains open as Madelaine and (perhaps even more to the point) Rosette forego their reading of gender and morality (at least as it is defined in nineteenth-century France) in favor of "pure" esthetic and erotic attraction for each other.

Gautier thus leaves us with some very real questions about several issues. The possibility of defining a sexual identity in any way other than anatomical, for example, is completely put into question; the traditional male subject/female object dichotomy, which one might expect such a novel to exploit, is rather subverted by the fact that both Madelaine and Rosette are infinitely more lucid, more observant, more literate in the language of bodies than d'Albert.[47] Indeed, d'Albert is a buffoonish figure when compared with the stronger, more capable women who surround him. Gautier, it would seem, ironizes the assumption of intellectual superiority and subjecthood solely on the basis of sex. Unlike many critics, who insist on seeing d'Albert as a sort of textual doppelgänger for Gautier, I would argue that, if such a reading must be done, Madelaine is the character who comes the closest to credibility and therefore the one whose pronouncements must be taken most seriously. In its own way, I think *Mademoiselle de Maupin* can be read as a feminist novel, complete with triumphant heroine riding off into the sunrise.[48] While d'Albert's discourse of woman-as-esthetic-object is firmly anchored in the persona of a foolish, inexperienced, somewhat dim-witted egomaniac, Madelaine's characterization of the subjugation of women in a system which assures their ignorance by categorically denying them access to knowledge is one of the very few things not ironized in the text. In fact, her quest for knowledge, of both herself and others, is ultimately a successful one, and it is by no means a feminist anachronism to see her as the true heroine of the novel.

As Rosemary Lloyd points out, the (revolutionary) questioning of gender roles is one of the salient aspects of Gautier's novel:

> Despite the pirouettes and jokes of *Mademoiselle de Maupin* . . . Gautier does have important points to make. Firstly, he emphasizes the confining nature of sexual stereotypes: if, in the wake of Laclos, he laments women's condition and education and reveals the emptiness of d'Albert's dream of happiness, it is above all to explain and deplore the fact that, as Madelaine puts it so succinctly, "il n'y a le moindre lien intellectuel entre les deux sexes."[49]

Indeed, in spite of its verbosity and confusion, *Mademoiselle de Maupin* is a novel which addresses some extremely important and provocative issues. Although the theme of reading the body in this text is of immediate and obvious concern to my study, its treatment leads to unexpectedly far-reaching and ambiguous problems whose interest far exceeds that of simple thematics and which join those seen in the other texts in this study. As we have seen, the question of determining literal, anatomical sex is fairly quickly put to use as an allegory of the question of defining the considerably more complex and ill-defined category of gender. We as readers know that Madelaine/Théodore is a woman, but we do not know what, if anything, that physical fact signifies; nor, ultimately, do the narrators.

Definitions of the metaphysical concept of gender always imply physiognomical assumptions, as they necessarily rely on belief in the existence of a direct correlation between corporeal and noncorporeal selves. Madelaine shows us that she is indeed a woman who possesses "masculine" attributes and the status of thinking, desiring subject they imply.[50] The fact that she must adopt a masculine mask in order to assume her true potential only serves to point out all the more clearly the fact that gender roles are merely social constructs. If Madelaine can be more of a "man" than some men (including d'Albert), how can the category of "man" used to denote anything other than anatomy be viable?

Gautier's insistence on leaving unanswerable questions of gender unanswered or, more exactly, on implying the infinite range of possible answers, provides us with an implicit but compelling critique of physiognomical thought. In contrast to more conventional thinkers, such as those we have seen in previous chapters (and indeed in contrast to the so-called protagonist of his novel), Gautier would seem to reject the notion of anatomy as destiny.[51] Madelaine de Maupin is ultimately a strong, intelligent, sexualized female character who subverts notions of gender essentialism. She is the negation of the other female characters in this study: neither a coquette like Marianne, a virginal monster like Mademoiselle Cormon, nor a femme fatale like Zola's Nana. The novel which bears her name, in contrast to the others with which I compare it in this study, functions as an exposition not of what bodies can signify, but rather of what they cannot signify.

The fact that the text itself is ultimately unable to answer the questions it poses, and that it avows this inability, is what distinguishes Gautier's novel as subversive with respect to traditions of physiognomical discourse in fictional narrative. *Mademoiselle de Maupin* performs the profoundly antiphysiognomical gesture of suggesting that the questions it has raised about the body are perhaps unanswerable; in so doing, it stands in direct contrast to the physiognomical novels of this study which tell us that answers exist without explaining how one might arrive at them. By thus putting into question the very notion of the legible body, it puts in relief the rhetorical and thematic phenomena common to the other works studied here. Ultimately, it serves as the proverbial exception that proves the rule.

It is important to note, however, that Gautier's modernity vis-à-vis the questions of sex, gender, and sexuality does not hail the beginning of a new era of literary texts which question these categories. On the contrary, Gautier's text represents a "blip" on the time line of the history of the treatment of sex, gender, and sexuality as literary themes. The second half of the nineteenth century in France, as is vividly evinced in its literary products, was a time of preoccupation with science and pseudoscience and their contributions toward a mystification of the female body. This period of genuine progress in scientific theory and method is characterized by increasingly paranoid attempts to define femininity within pseudoscientific discourse as the locus of disease and infection. The physical (corporeal) and the metaphysical (moral) were of course inseparable in a totalizing and essentialist rhetoric, in which once again the body is pressed into service as a "scientific" signifier.

Zola's *Nana*, as perhaps the most extreme literary exemplar of the seemingly limitless misogyny of the late nineteenth century, will therefore bear no resemblance to *Mademoiselle de Maupin*. Indeed, the ideological implications of the two texts are at opposite poles with respect to the question of the signifying body. Whereas Gautier seems to celebrate androgyny, ambiguity, and desire, Zola represents a morbid fear of them. If Gautier can be said to prefigure a twentieth-century avowal of the unanswerability of certain questions (particularly those posed by the human body), his successor Zola represents a remarkably consistent late-nineteenth-century insistence on imposing systems of artificial meaning by which those very questions can be answered.

Chapter Six

Zola and *le signe de la femme fatale*

Nana était nue. Elle était nue avec une tranquille audace, certaine de la toute-puissance de sa chair . . . Tout d'un coup, dans la bonne enfant, la femme se dressait, inquiétante, apportant le coup de folie de son sexe, ouvrant l'inconnu du désir. Nana souriait toujours, mais d'un sourire aigu de mangeuse d'hommes.

—Zola

Unfortunately for my purposes, Zola has left virtually nothing on the subject of physiognomy in his extensive critical and theoretical writings, and his novels, particularly those of the *Rougon-Macquart* cycle, do not seem to emphasize physiognomical analysis of characters. The relation between Zola and physiognomy is neither explicit nor obvious, as was the case for example with Balzac. Indeed, by the late nineteenth century, physiognomy had lost most (if not all) of its former acceptance as a science, as the discourse and methodology of the sciences became increasingly positivistic. What is perhaps most compelling for me is the fact that the apparent death of physiognomy did not signal the death of physiognomical concerns in literature; Zola's work provides us with an especially acute and intriguing illustration of this fact.

The concerns of this chapter will therefore be less literary-historical than analytic, a study of those aspects of Zola's work which are implicitly linked to physiognomical thought: treatment of physical portraiture, the function of the body in narrative, the real and imagined links between literature and science, and fields of speculation in which the physical and the moral are seen to be inextricably intertwined.

There is, nonetheless, at least one article in the vast body of criticism on Zola which explicitly treats the relation of Zola to physiognomy per se: E. P. Gauthier's "New Light on Zola and Physiognomy," which appeared in 1960. Although brief and not especially insightful, this article, as the only one which directly addresses the topic, provides an appropriate starting point for my discussion. Gauthier points out illustrations in Zola of certain traditional principles of physiognomy, including "the most fantastic theory of certain physiognomists; the claim that human resemblances to various animals are highly significant indices to the recognition of character,"[1] examples of which he finds in *La Bête humaine* and *La Fortune des Rougon*. Gauthier goes on to mention that Prosper Lucas, a theorist of heredity and a widely recognized influence on Zola, was rather skeptical in regard to physiognomy, as were most of his scientific contemporaries, but does mention Lavater in the introduction to his *Traité philosophique et physiologique de l'hérédité naturelle* (1850).[2] Gauthier also tells us that Balzac used his edition of Lavater "liberally, and acknowleged his debt." Gauthier's ultimate thesis is in fact that Zola "found it no less precious [than did Balzac], used it almost as liberally, but did not credit his source."[3] Indeed it was Zola's reading of Balzac that led him to an interest in physiognomy, states Gauthier. Although this assertion, and indeed the entire thesis about Zola's use of physiognomy, is probably an exaggeration, it does suggest the fundamental, if not novel, idea that it was through Balzac that Zola came to conceptualize the notion of fusing "science" and fiction in a work of literature.

The specifics of the influence of the *Comédie humaine* on the *Rougon-Macquart* cycle are less pertinent here than the basic conceptual analogies to be drawn between Balzac's means of analyzing and systematizing human life and those practiced by Zola. As we saw in a preceding chapter, Balzac's central preoccupation would seem to be a search for explanations of human life, for validation of systems of interpretation which specifically use the physical world as an index for the metaphysical, the visible for the invisible. For Balzac, physiognomy was precisely such a system, and therefore of invaluable use to his project. The fundamental agenda of Zola's ambitious *Rougon-Macquart* cycle (significantly subtitled *Histoire naturelle et sociale d'une famille sous le second empire*) is also that of uncovering systems of explanation, both "natural" and social, of human behavior. Although Zola posits as his authority different fields of "scientific inquiry" than does Balzac, his basic agenda is the same.

Zola was directly influenced by fields of study more topical in the late nineteenth century than physiognomy, namely heredity, experimental medicine, and public health.[4] Zola's pertinence to our study comes from the fact that these areas of concern share the questions posed earlier by physiognomy, and indeed have in common with physiognomy the goal of ordering our knowledge of life through correlations between that which is immediately accessible to us (present in either a physical or a temporal sense, the human body being a privileged example) and that which is beyond the immediate field of our vision or experience. The more naïve assertions and methods proposed by physiognomy were replaced in Zola's time with more sophisticated hypotheses with a closer, if still tenuous, relation to "real" science. Zola's work, therefore, reveals its truest relevance to our discussion of physiognomy and literature when read less as yet another example of fictional discourse demonstrating physiognomical preoccupations than as the conceptual and historical product of such preoccupations. That is to say, the concerns of physiognomy remain in a general sense intact in Zola, while at the same time the modes of "investigation" change dramatically. We therefore necessarily face the problem of distinguishing in Zola's work between attempts at anthropological interpretation which are a legacy from earlier systems and those which reveal a genuinely novel progression of thought. Such a reading of Zola should provide valuable insight into both Zola's work and the larger import of these concerns so prevalent in works of narrative literature.

Le Roman expérimental

Perhaps the most logical way to analyze the relation between literature and science in Zola would be to look first at his treatise on the subject, *Le Roman expérimental*. First published in 1879 in the same issues (16–20 October) of the journal *Le Voltaire* as the first installments of a serialized *Nana*, *Le Roman expérimental* is the manifesto of naturalism.[5] In it, Zola articulates an analogy between the work of the scientist and that of the novelist; their methods, he tells us, should be the same. Like the scientist, the novelist needs to rely on case studies, observation, and other positivistic data in order to develop a hypothesis of some kind about the probable result of placing certain variables (in the novelist's case, characters) in certain environments. This stage is *observation,* and is linked to the next and more important stage, *expérimentation* (whence the title of the treatise). In the stage of experimentation, the scientist actually provokes the hypothetical circumstance and records the results. If the observation phase might seem to provide a possibly justifiable analogy between the work of a scientist and that of a novelist, the analogy of the experimentation phase seems to require a suspension of disbelief that few other than Zola himself have been able to achieve.

Perhaps the problem with *Le Roman expérimental* is that Zola has taken his project, that of transposing the then-radical theories of experimental physiology proposed by Claude Bernard in his *Introduction a l'étude de la médecine expérimentale* (1865), much too literally. Indeed, the method of observation and experimentation comes directly from Bernard, and there are instances in which Zola seems to have taken passages from Bernard and merely substituted a few words here and there, with unsatisfactory results. To be sure, Zola does not hide this fact, calling his treatise a "travail d'adaptation" and a "compilation de textes." In fact, Zola states on the very first page that "Le plus souvent, il me suffira de remplacer le mot 'médecin' par le mot 'romancier,' pour rendre ma pensée claire et lui apporter la rigueur d'une vérité scientifique."[6]

Before continuing with any commentary about *Le Roman expérimental*, it would perhaps be most useful to allow the text to define itself in a few key passages:

> En revenant au roman, nous voyons également que le romancier est fait d'un observateur et d'un expérimentateur. L'observateur chez lui donne les faits tels qu'il les a observés, pose le point de départ, établit le terrain solide sur lequel vont marcher les personnages et se développer les phénomènes. Puis, l'expérimentateur paraît et institue l'expérience, je veux dire fait mouvoir les personnages dans une histoire particulière, pour y montrer que la succession des faits y sera telle que l'exige le déterminisme des phénomènes mis à l'étude. C'est presque toujours ici une expérience "pour voir," comme l'appelle Claude Bernard. *Le romancier part à la recherche d'une vérité.* (63–64; emphasis mine)

> Si le romancier expérimental marche encore à tâtons dans la plus obscure et la plus complexe des sciences, cela n'empêche pas cette science d'exister. Il est indéniable que le roman naturaliste, tel que nous le comprenons à cette heure, est une expérience véritable que le romancier fait sur l'homme, en s'aidant de l'observation. (64–65)

> Au lieu d'enfermer le romancier dans des liens étroits, la méthode expérimentale le laisse à toute son intelligence de penseur et à tout son génie de créateur. Il lui faudra voir, comprendre, inventer. Un fait observé devra faire jaillir l'idée de l'expérience à instituer, du roman à écrire, pour arriver à la connaissance complète d'une vérité. (66)

> Je résume . . . en répétant que les romanciers naturalistes observent et expérimentent, et que toute leur besogne naît du doute où ils se placent en face des vérités mal connues, des phénomènes inexpliqués, jusqu'à ce qu'une idée expérimentale éveille brusquement un jour leur génie et les pousse à instituer une expérience, pour *analyser les faits et s'en rendre les maîtres.* (67; emphasis mine)

The passages quoted above give a fairly clear idea of the premises of *Le Roman expérimental*, and thereby demonstrate several of the salient points where Zola's theoretical claims simply do not withstand scrutiny. Perhaps the most important of these points is revealed by Zola's liberal, if not to say erroneous, use of the term *faits*. When he tells us that the work of the novelist begins with "les faits tels qu'il les a observés," he would have us believe that an "experimental" novel is the fruit of extensive and specific fieldwork, a sort of sociological case study. We know, from the extensive documentation Zola compiled concerning the preparation of his novels, that firsthand "investigation" was in fact a part of his process of *observation*, but we know as well that much of it came from second- or thirdhand anecdotes and books about whichever sector of society he was treating at the moment.[7] The "facts" to which he refers are much more likely to be myths or generalities than the result of "scientific" observation of specific individuals.

The move to the next step, that of experimentation, reveals an even more dubious use of the term *faits*. Zola claims that the novelist "fait mouvoir les personnages dans une histoire particulière" in order to demonstrate that the resultant *faits* will be in accordance with the *déterminisme* of the phenomena being studied. The definition of this *déterminisme* is, however, the result of the *faits* of the observation stage and must be at best a personal hypothesis on the part of the novelist based on perhaps, as is the case for Zola, questionable data. The novelist supposedly takes this hypothesis and "proves" it by the actions of the characters in his story. The succession of *faits* in the story validates, Zola would have us believe, the author's hypothesis. Indeed, it should come as no surprise that the "facts" of a novelist's narrative should bear out his hypothesis, since they are both of his creation and subject to no factual verification whatsoever. Zola tells us that "un fait observé" can "faire jaillir l'idée de l'expérience à instituer, du roman à écrire," suggesting that writing a fictional narrative is in some way equivalent to conducting a laboratory experiment; it is the logic of this equivalence that remains unarticulated by Zola and inaccessible to the reader of his theory.

It remains unclear throughout *Le Roman expérimental* if Zola recognizes the most fundamental problem in his theory, that the specifics of a fictional narrative can in no way be characterized as fact, and are therefore of no value whatsoever in proving any kind of extratextual theory. Zola posits a hypothesis based on semifact in order to have it proved by complete fiction—the whole process to be accepted as scientific, as equivalent to the most positivistic of methods. It is indeed ironic that Zola chooses as his model the observation/experimentation method as proposed by Bernard, a branch of scientific thought that dogmatically emphasizes reliance on nothing but empirical evidence in order to validate hypotheses.

Zola recognizes at various moments in his text that the experimental novel is the youngest "science," and that it does in fact retain imaginative elements which have yet to be proved; however, these disclaimers, even if protected by quotes from Bernard, are insufficient rhetorical defenses of his theory. What remains constant is the assertion on Zola's part that the novel, by definition fictional discourse, is an appropriate forum for proving psychosociological, even biological, theories. It is important for poststructuralist readers to bear in mind that Zola is not referring to proof on an intratextual level, the bearing out of the internal coherence of a work of literature; he is, in no uncertain terms, claiming that fiction can prove hypotheses about real, extratextual human phenomena in the same way that laboratory experiments can. In so doing, he posits fictional narrative as fact. Even the most determined poststructuralist would have a hard time establishing that the difference between scientific and fictional discourse is merely an arbitrary generic one; there exist very real and undeniable differences between proof realized in the physical and material world of the laboratory and that which is realized in the imaginative and linguistic constructs of a novelist. The equation of scientific method and the novelistic process, the core of Zola's argument, does not hold up to any critical scrutiny whatsoever, as he simply fails to articulate a credible analogy between the two enterprises.

In addition to the egregious conceptual problem which undermines the very project of Zola's text, there are other aspects of his work which retain my attention for the purposes of this study. It may be of some use to my discussion to point out several clear analogies to be drawn between Zola's theory of the experimental novel and physiognomy, on both the methodological and the rhetorical planes.

First, Zola uses the concept of genius to justify the often mind-boggling leaps of imagination necessary to move from the stage of observation to that of experimentation. He tells us that the experimental method allows the novelist to rely on "tout son génie de créateur" (66). As we have seen in the previous chapters on Lavater and Balzac, the idea of genius, of innate powers of perception and interpretation, is one which seems to be inextricably linked to physiognomical thought; here we find that it is an integral element in the theory of the experimental method as well. Zola tells us that the novelist must employ a scientific mode in the gathering of facts, which are then submitted to his genius for processing into a narrative. However, the genius required in both physiognomy and Zola's version of experimentation is that of imagination, a step in the method which cannot be demonstrated, proved, or disproved and is therefore in direct contradiction with any definition of scientific method. As I stated above, this might not be problematic if the goal were to create a work of literature with its own rhetorical coherence and without claims to scientific veracity; Zola's goal is, however, to create a work of credible scientific discourse. Indeed, he states the goal of his work in terms free of ambiguity and, incidentally, reminiscent of Lavater: "Le romancier part à la recherche d'une vérité" (64); "pour arriver à la connaissance complète d'une vérité" (66); "pour analyser les faits et s'en rendre les maîtres" (67).

In both its means and its objectives, Zola's treatise parallels physiognomical thought in that, for both, the process is as follows: the gathering of roughly factual data in regard to which a fictional narrative is created which in turn supposedly reveals certain truths about the data and their meaning in the world. The most important link between the two methods lies in the fact that each is based on the creation of a more or less fictional narrative of explanation which is posited as scientific discourse.[8] The line between imagination and reality, admittedly difficult to draw, becomes completely obliterated as speculation becomes "scientific" certainty. This rhetorical transsubstantiation takes place with no explanation or justification offered to the bewildered reader; it requires a suspension of disbelief much more appropriately demanded of one of the faithful at the celebration of a religious rite than of a reader of scientific theory.

In spite of the fact that Zola's equation of the novel with science remains unconvincing, it does point up in the process an extremely important link between the two enterprises: narrative discourse. In an insightful and informed article on the analogy between plot and science in late-nineteenth-century literature, Gillian Beer suggests that fictional narrative and scientific discourse have the common agenda of attempting to define meaning through systems, through the study of relations. Perhaps no author demonstrates these preoccupations more clearly than Zola. Beer also defines plot and its relation to scientific method, and in so doing, outlines the central issue of the self-verifying function of fictional narrative I discussed above in Zola:

The belief that fixed natural laws underlie process could provide a source of authority for narrative . . . In the nineteenth century the novelists' high value for individuality tended always toward the perception of divergencies, but the power to perceive *systems* expressed through individuals was shared by scientists and novelists alike and this stabilizing power was a deep organizing principle for their fiction . . . The *methods* of scientists become the methods of emplotment and scientific theories suggest new organizations for fiction.

Plot in nineteenth-century fiction is a radical form of interpretation: it fixes the relations between phenomena. It projects the future and then gives real form to its own predictions. Thus, it is self-verifying: its solutions confirm the validity of the clues proposed. Such plot assumes that what is hidden may be uncovered, and that what lies beyond the peripheries of present knowledge may be encompassed and brought within the account by its completion. In this sense it shares the nature of hypothesis, which by its causal narrative seeks ultimately to convert its own status from that of idea to truth.[9]

Beer articulates in the passages above the key relation between novelistic plot and science: that of the creation or perception of systems through narrative discourse. This is a relation only hinted at by Zola in *Le Roman expérimental*, but one which is fundamental to the logic of his work.

At the risk of being accused of positivism myself, I must reiterate here that the distinction to be made between the two enterprises is that the scientific method is, more often than not, the perception of systems which have some kind of verifiable existence in nature; the novel, on the other hand, is more often than not the creation of imaginary systems which explain patterns of human existence. To be sure, it can be said that both scientific and novelistic systems are creations of human consciousness, and can be expressed only through linguistic constructs. However, I would maintain that there is at the very least a smaller part of speculation and a greater part of indisputable empirical fact to be found in most scientific discourse than in most novelistic discourse (including that of Zola), and that this is necessarily true by the very nature of the two disciplines.

This distinction notwithstanding, the overriding commonality as defined by Beer is that of the process of ordering, a fundamental element of traditional narrative coherence or plot. As we saw earlier in discussions of Peter Brooks's definition of plot, the most basic function of traditional narrative can be said to be the ordering (through the medium of language) of phenomena in sequential, logical fashion, thereby producing coherence and system.[10]

It is the will to order, through coherent, causal narrative, that links science and fiction and so completely seduces Zola. As Beer suggests above, it is in this way that science and literature have their most profound mutual influence. As we shall see in a discussion of Zola's novels below, nothing is more central to his project than the desire to create a system of coherence. It is important to note in passing that this desire (the "will to plot," if you will) reaches its peak in the late nineteenth century in both science and literature, with Zola being perhaps the most exemplary novelist and Darwin the most exemplary scientist.[11]

It is perhaps not surprising that Zola's theory of the experimental novel is contradicted by the ethos of his novels. In spite of his undoubtedly sincere theoretical beliefs, the novels would seem to betray obsessions and personal mythologies that soon escape his "scientific" principles.[12] As I mentioned above, Zola's investigation for the preparation of his novels probably remained fairly superficial by scientific standards (in spite of his visits to coal mines, hospitals, and morgues), and therefore both his hypotheses and their "proof" reveal much more about Zola than, as he would have it, about any extratextual biological or social realities.

Nowhere is the *écart* between Zola's positivistic principles and his lurid imagination more evident than in *Nana*. Beginning with Flaubert's much-quoted "Nana tourne au mythe . . .," there is a long tradition among readers of the novel to expose the "mythological" structures underlying *Nana*. Indeed, the novel is much more accurately defined by its "mythological," even fantastic content than by any experimental or naturalistic procedures evident in the text. It has often been speculated, perhaps not without reason, that Zola unwittingly reveals some of the darker and less healthy of his fantasms in *Nana*. While I am reluctant to engage in a posthumous psychoanalysis of the novelist, I cannot argue with the prevalence and significance of the myths in this most disturbing of Zola's works.

The central myth in *Nana*, the myth to which all others are in some way subordinate, is that of the *femme fatale*.[13] Contrary to Zola's stated naturalist agenda, the title character in his novel is much more the luridly caricatural illustration of sexualized woman as agent of death and destruction than the "realistic" portrayal of an exemplar of a sociological or biological type.[14] *Nana* is more pathological fantasy than case study, and its content cannot be said to prove any hypothesis, scientific or other.

This is not to say, however, that Zola abandons his concerns with sociology, biology, and heredity in *Nana*. On the contrary, in certain key passages, he makes rather a point of Nana's tainted sociobiological heritage and consequent inherent predisposition to vice. Nonetheless, as I stated above, it is less the completely predictable mise-en-scène of Zola's scientific theories through his fiction than the mythological content of his "science" as expressed in both fiction and theory that interests me here.[15] As my specific interest in all of this is the role of the body and particularly the body as potential sign, *Nana* provides the most compelling object of analysis of all of Zola's novels. *Nana* is indeed nothing if not the ultimate myth of the body, and specifically that of the female body. My discussion of the body in this novel will necessarily entail discussion of Zola's most central scientific theory, that of heredity, and of the ideology of biological and sociological determinism his version of it implies.

Nana

Nana might seem at first an odd choice as an object of analysis of physiognomy and literature: it contains very little of what can be called physiognomical description, even of its title character. Nowhere in the course of his narrative does Zola express any explicitly physiognomical concepts, and few of the characters in the novel engage in the act of reading the body. However, the centrality of Nana's body to the narrative, and the thus curious absence of description of the usual physiognomically significant features (the facial features, for example), render *Nana* a fertile ground for exploring the very conditions of possibility of physiognomical pursuits. The novel may indeed be read as a problematization of the act of physiognomical reading.

The first thing one notices about *Nana* with regard to physiognomy is that Nana is much less an object of physiognomical analysis than a subject who provokes physiognomical reactions in others. The narrative hinges on the desire the *demi-mondaine* sparks in men, and this very physical desire is often betrayed by physiognomical (or, to use Lavater's term, pathognomical) indices. The text therefore, in keeping perhaps with Zola's stated agenda of writing "le poème du désir du mâle" in *Nana*, offers men in a state of sexual hyperexcitement as objects of physiognomical interest. The examples are numerous and the descriptions remain fairly similar from one instance to another:

> D'un mouvement, elle s'était penchée, ne s'étudiant plus; et son peignoir
> ouvert laissa voir son cou, tandis que ses genoux tendus dessinaient,

sous la mince étoffe, la rondeur de la cuisse. Un peu de sang parut
aux joues terreuses du marquis. Le comte Muffat qui allait parler, baissa
les yeux.[16]

[Nana in reference to her crowd of avid suitors]
Ils devaient avoir une bonne tête, tous la langue pendante . . . (79)

[at the very mention of Nana's name]
Le comte devint plus grave. Il eut à peine un battement de paupières,
pendant qu'un malaise, comme une ombre de migraine, passait sur son
front. (99)

[on watching Nana undress]
Le comte s'approcha de la psyché, se vit très rouge, de fines gouttes de
sueur au front; il baissa les yeux . . . il s'assit au bord du divan capitonné,
entre les deux fenêtres. Mais il se releva tout de suite, retourna près de
la toilette, ne regarda plus rien, les yeux vagues . . . (151)

Le comte Muffat venait tous les soirs, et s'en retournait, la face gonflée,
les mains brûlantes. (190)

Georges regardait Nana avec un tel bonheur . . . que ses yeux s'emplis-
saient de larmes. (302)

Lui, restait le coeur serré, n'osant plus bouger, ayant des rougeurs de
fille, aux moindres mots. (306)

It is to be noted in passing that most of the examples above concern
the comte Muffat, Nana's most obsessive and persistent suitor. The
comte serves as the most extreme example of the disquietude, guilt,
and general confusion that intense sexual desire causes in *Nana;* he is,
however, only the clearest example of this condition and not essentially different from any other male character aroused by Nana's body
in the novel.

Although Zola does not emphasize the point, it is also important
to realize that Nana is, at least to some extent, aware of the physiognomical legibility of the subjugation to which she condemns those
who desire her:

Une fois encore, Steiner était pris, et si rudement que, près de Nana, il
restait comme assommé, mangeant sans faim, la lèvre pendante, la face
marbrée de taches. Elle n'avait qu'à dire un chiffre. Pourtant, elle ne se
pressait pas, jouant avec lui, soufflant des rires dans son oreille velue,
s'amusant des frissons qui passaient sur son épaisse figure. (121)

Elle répondait par de petits mouvements de tête, tout en faisant de rapides réflexions. Ça devait être le vieux qui avait amené l'autre; ses yeux étaient trop polissons. Pourtant, il fallait aussi se méfier de l'autre, dont les tempes se gonflaient drôlement; il aurait bien pu venir tout seul. (74)

Contrary to traditional gender roles (and some would say that this subversion is precisely the danger Nana represents), in *Nana* it is the woman who performs the act of physiognomical reading and is therefore in the privileged and powerful position of knowing spectator. Nana is the object of desire, and it is indeed thus that she can become the reading subject of physiognomy.

Nana is introduced in all of her sexuality and corporeality to both the characters and the reader of the novel in a very literal mise-en-scène of the dynamic of the entire novel: she appears, quasi-nude, on stage before a crowd of confused, desirous, red-faced men.[17] This opening scene of the novel encapsulates the basic premise of the entire text, as Nana's body provokes desire so extreme as to invite confusion and even chaos in a group of men while she herself remains impassive. Zola's description of Nana's carnality and its devastating power over men prefigures the entire narrative:

Un frisson remua la salle. Nana était nue. Elle était nue avec une tranquille audace, certaine de la toute-puissance de sa chair. Une simple gaze l'enveloppait; ses épaules rondes, sa gorge d'amazone dont les pointes roses se tenaient levées et rigides comme des lances, ses larges hanches qui roulaient dans un balancement voluptueux, ses cuisses de blonde grasse, tout son corps se devinait, se voyait sous le tissu léger, d'une blancheur d'écume . . . Et lorsque Nana levait les bras, on apercevait, aux feux de la rampe, les poils d'or de ses aisselles. Il n'y eut pas d'applaudissements. Personne ne riait plus, les faces des hommes, sérieuses, se tendaient, avec le nez aminci, la bouche irritée et sans salive. Un vent semblait avoir passé, très doux, chargé d'une sourde menace. Tout d'un coup, dans la bonne enfant, la femme se dressait, inquiétante, apportant le coup de folie de son sexe, ouvrant l'inconnu du désir. Nana souriait toujours, mais d'un sourire aigu de mangeuse d'hommes. (53)

Among other things, the passage above, perhaps the most significant of the novel, demonstrates the manner in which Nana's body will be described throughout, in terms of sheer corporeality, as a well-shaped, fragrant mass of white flesh. The reader sees Nana through the eyes of the male characters whom she dazzles; their perception, and by consequence ours, is an impressionistic one, of a figure of shape ("ses épaules rondes"), of color ("les pointes roses," "d'une blancheur d'écume," "les poils d'or"), and of movement ("ses larges hanches qui roulaient dans un balancement voluptueux").[18] It is not, however, a perception which includes specific facial features or other corporeal details of potential significance to a physiognomical reading. It is furthermore neither the distance from the stage nor the effects of theatrical lighting which create this soft-focus image, for we see the same kind of fleshy, indiscriminate image in all descriptions of Nana's body.

In each case, we see Nana through the eyes of a male spectator character whose view of Nana is of soft, white flesh covered by golden hair.[19] These male characters, overcome in each case by extreme sexual arousal, see her body simply as a tool for their own pleasure, as a material object. It is thus easy to understand why the act of physiognomical reading would not come into play: because of their desire, the men cease to have any sense of Nana as a human being, as a composite of physical and moral selves, let alone as a complex amalgamation of potential signs. This is indeed at the very core of Zola's agenda for the novel, as he states in no uncertain terms in his *ébauche* of *Nana:* "Elle est la chair centrale . . . Ne pas la faire spirituelle, ce qui serait une faute; elle n'est que la chair."[20] Having lost a sense of the two-sided human nature which defines the physiognomical enterprise, male characters see Nana in terms of pure materiality and fail to recognize the need to read beyond her secondary sexual characteristics. They fail to read beyond their own desire.

Janet Beizer states the problem in what is perhaps the most important discussion to date of the problematic of body-as-text in *Nana:*

> her [Nana's] body becomes text, taking the place of the lines she does not deliver and the songs she cannot sing . . .
> We have an initial displacement of the signifier from language to the body—word made flesh. As the novel progresses, we in fact find the locus of meaning repeatedly relayed from the word to the body.[21]

Zola suggests, albeit subtly, that there is indeed something in Nana to be read. At the end of the lengthy passage from *Nana* cited above, he tells us that suddenly the woman in Nana came forth ("la femme se dressait"), bringing with her the "coup de folie de son sexe." It is important to see here that the "coup de folie" is not her own insanity, but rather the insanity she communicates to men; her "sexe," on the other hand, refers both to her gender, and, quite literally, to her genitals.

It is precisely this "coup de folie," a blindness brought on by sexual excitation to the point of distraction, that prevents the men in the audience from seeing Nana's smile or recognizing its significance, the "sourire aigu de mangeuse d'hommes." If the male characters were not dominated by their physical desire, it is suggested, they might recognize the danger Nana represents.[22] Of course, the irony of the novel (and the very definition of the femme fatale) is that by the time they are close enough to distinguish the evil, they have already been blinded by desire and are beyond the point of no return, headed for a vertiginous fall into Nana's voluptuous abyss. These ideas, that of the blindness of desire and the unseen evil and danger of woman, are the myths which fuel Zola's narrative. The novel, as chronicle of this blindness and the fall it provokes, serves to demonstrate these myths in efficient and vivid fashion.

The sexual desire Nana provokes in men with such consistency is perhaps due to the fact that they, and we, so frequently see her nude (or at least seminude). There is, after all, a certain logic to the idea that one might be so distracted by the sight of a nude body as to fail to distinguish potentially legible corporeal details. Nana's powers of attraction are clearly those of the fascination of simple carnality, the material reality of the human body:

> Tous se tournèrent. Elle ne s'était pas couverte du tout, elle venait simple-
> ment de boutonner un petit corsage de percale, qui lui cachait à demi la
> gorge. Lorsque ces messieurs l'avaient mise en fuite, elle se déshabillait
> à peine . . . Par derrière, son pantalon laissait passer encore un bout de
> sa chemise. Et les bras nus, les épaules nues, la pointe des seins à l'air,
> dans son adorable jeunesse de blonde grasse, elle tenait le rideau d'une
> main, comme pour le tirer de nouveau, au moindre effarouchement . . .
> tranquillement, pour aller à la toilette, elle passa en pantalon au milieu
> de ces messieurs, qui s'écartèrent. Elle avait les hanches très fortes, le
> pantalon ballonait, pendant que, la poitrine en avant, elle saluait encore
> avec son fin sourire. (151–52)

> Au milieu du grand silence, un soupir profond, une lointaine rumeur
> de foule, montait. Chaque soir, le même effet se produisait à l'entrée
> de Vénus, dans sa nudité de déesse. (163)

> Alors, il [Muffat] leva les yeux. Nana s'était absorbée dans son ravisse-
> ment d'elle-même . . . Lentement, elle ouvrit les bras pour développer
> son torse de Vénus grasse, elle ploya la taille, s'examinant de dos et
> de face, s'arrêtant au profil de sa gorge aux rondeurs fuyantes de ses
> cuisses . . . ne pouvant détourner les yeux, il la regarda fixement, il tâ-
> chait de s'emplir du dégoût de sa nudité . . . Il avait conscience de sa

défaite, il la savait stupide, ordurière et menteuse, et il la voulait, même empoisonnée. (216–17)

Zoé introduisait un monsieur tout tremblant dans le cabinet de toilette, où Nana changeait de linge . . . C'était Georges, en effet. Mais, en la voyant en chemise, avec ses cheveux d'or sur ses épaules nues, il s'était jeté à son cou, l'avait prise et la baisait partout. (302)

Indeed, one can easily follow the Zolian logic by which a heterosexual man might be distracted by his senses in the presence of a beautiful, quasi-nude female body. In his own lurid fashion, Zola describes not only the sight of Nana's body, but also, at least implicitly, its feel, taste, and especially smell:

il prit les cinquante francs; mais une pièce resta, et il dut, pour l'avoir, la ramasser sur la peau même de la jeune femme, une peau tiède et souple qui lui laissa un frisson. (75)

Le champagne qu'elle avait bu la faisait toute rose, la bouche humide . . . Il voyait là, près de l'oreille, un petit coin délicat, un satin qui le rendait fou. (125)

Le rut qui montait d'elle, ainsi que d'une bête en folie, s'était épandu toujours davantage, emplissant la salle. (54)

craignant de défaillir dans cette odeur de femme qu'il retrouvait . . . (151)

The examples above are but a few of the recurrent descriptions of the power of Nana's sensual attractions and the distraction they cause in the men who come into contact with her.[23] Nana is, above all, a *bête*.[24] Portrayed as pure soulless carnality, she provokes simple mindless desire in others, thus degrading them to the level of *bête*. This, Zola would seem to suggest, is the danger of Nana: by reducing men to the level of their senses, she robs them of their powers of reason and judgment. For Zola, would-be *homme de science* and veritable fetishizer of observation, experimentation, and method, this is obviously the central point of her powers of destruction and of the novel itself.

In *L'Eros et la femme chez Zola*, Chantal Bertrand-Jennings gives a convincing and concise reading of the significance of Nana as *bête*, exemplified by the scene at the racetrack in which a horse named for Nana wins the race:

au champ de course, il n'est plus possible de la distinguer de la pouliche du même nom. Ce n'est donc pas par hasard si "Nana" va battre la pou-

liche anglaise "Spirit." C'est que la chair féminine . . . est enfin parvenue
à vaincre l'Esprit. De plus, pareille à une nouvelle Circé, Nana possède
le don de révéler la bestialité sous-jacente des individus distingués qui,
en l'approchant, se métamorphosent en une "meute d'hommes," ou en
un "troupeau" galopant à travers son alcôve.[25]

If Nana's body serves to destroy men's ability to discern her potentially significant features (her "mangeuse d'hommes" smile, for example), is it not possible that, paradoxically, her nudity acts as a
mask?[26] If it is her nudity which creates desire, which in turn blinds
the desiring subject through a sensual confusion, does her nudity
not act as an obstacle to "true" sight? In this sense, I would argue
that Nana's body as material object obliterates itself as signifying object. It would seem that, in this novel at least, these two primary
functions of the body, that of object of analysis and that of object
of desire, are mutually exclusive. Desire necessarily precludes reason
and even sight.[27]

On an even more paradoxical level, and in contrast to traditional
physiognomical concepts, we can say that Nana's beauty produces an
effect which is in direct contradiction with the "truth" of her inner
being; physiognomical reading would have to become, therefore, a
question of reading against the grain of seemingly transparent signs.
In other words, in Nana's case, the act of physiognomy would have to
be defined by a reading which would directly reverse the traditional
and literal correspondence between physical and metaphysical traits.
Nana's beauty would have to be recognized as indicative of evil, her
rosy health as disease, and her carnality as a sign of death. Nana's
body would have to be read as a mask or a shield of her inner being,
rather than a mirror of it. This, obviously, is a reading which will never
take place. Even if the femme fatale did not render men incapable
of reading her body, the perversity of the system of inverse correspondence between Nana's healthy, desirable body and her destruc-
tive, diseased essence would render physiognomical reading virtually
impossible. The few physiognomical clues Zola does provide about
Nana either are at moments when her sexual appeal has overwhelmed
and blinded her spectator(s) (as in the case of the smile above) or
are so obscure as to be beyond recognition (her mole, which I will
discuss below).

Indeed, *Nana* seems to illustrate the impossibility of reading the
body. The "message" of the novel with regard to physiognomy is perhaps not so much that there is no relationship between the body and
an essential inner being as that such a relationship is, at least in some

cases, beyond usual powers of perception and conceptualization. In this sense, one might accurately say that the message of impossibility (or at least limitation) of *Nana* is in direct contradiction to Zola's positivistic theory of the experimental novel.

It is intriguing that, perhaps in reaction to the message of illegibility in *Nana*, Zola reverses the phenomenon at the end of the novel, with Nana's death. Indeed, Nana's death turns her body into a very literal representation of the horrors of her soul. When she is stricken with smallpox, the mask of her beauty is finally removed and moral putrefaction is at last translated into corporeal putrefaction:

> C'était un charnier, un tas d'humeur et de sang, une pelletée de chair corrompue, jetée là, sur un coussin. Les pustules avaient envahi la figure entière, un bouton touchant l'autre; et, flétries, affaissées, d'un aspect grisâtre de boue, elles semblaient déjà une moisissure de terre, sur cette bouillie informe, où l'on ne retrouvait plus les traits. Un oeil, celui de gauche, avait complètement sombré dans le bouillonnement de la purulence; l'autre à demi ouvert, s'enfonçait, comme un trou noir et gâté. Le nez suppurait encore. Toute une croûte rougâtre partait d'une joue, envahissait la bouche, qu'elle tirait dans une rire abominable. Et, sur ce masque horrible et grotesque du néant, les cheveux, les beaux cheveux, gardant leur flambée de soleil, coulaient dans un ruissellement d'or. Vénus se décomposait. Il semblait que le virus pris par elle dans les ruisseaux, sur les charognes tolérées, ce ferment dont elle avait empoisonné un peuple, venait de lui remonter au visage et l'avait pourri. (438–39)

It is important to note that the horrifying passage above constitutes the longest and most detailed description of Nana's face in the novel as well as the final page of *Nana*. It is only after a long and suspenseful buildup in the death chamber that we are given the hideous description above. It is as if Zola's horrific monster story achieves its ultimate goal through the description of corporeal horror. Nana, agent of putrefaction and death, finally metamorphoses into her true form.

All ambiguity and illegibility are eradicated in the scene above, along with all possibility of desire for Nana's body.[28] Beizer gives Nana's disfigurement the opposite interpretation: "Her death in a sense effaces the conflict [of the novel], for it renders the 'page' illegible."[29] I would argue, however, that the conflict of the novel can be defined as the tension created by the contradiction between Nana's beauty and her powers of evil. In that sense, the conflict is indeed effaced by her disfigurement, as it is her mask, her *signes trompeurs*, which are obliterated by the disease, and her essential self which becomes undeniably visible. This effacement of the mask of her beauty is the condition for the literalization of her wickedness.[30]

Indeed, the physical "signs" of Nana's moral monstrosity are so literal as not to be signs at all; it is a question not of interpreting Nana's face (whose traits are, as Beizer points out, no longer recognizable), but of simply viewing its unambiguous monstrosity.[31] Zola suggests not that the rotting body *symbolizes* the rotten soul, but rather that the physical and metaphysical are at last one. There can no longer be any contradiction between the two sets of traits, as there is no longer any distinction between the two domains.

If, until her death, Nana's body provides only deceptive signs with regard to her inner being, the novel is not without explicit mention of her inalterably corrupt nature. The episode in the novel which most clearly points out and indeed explains Nana's depravity is the journalist Fauchery's allegorical profile of Nana as "la Mouche d'or." In this widely commented text within a text, Fauchery outlines Nana's origins and rise to power. Nowhere is Zola's fear of and repulsion for Nana more explicitly stated; indeed the piece can be read as a sort of allegory of the entire narrative of *Nana*. At the very least, it provides some significant keys to an understanding of the novel and its premises.

Fauchery's "chronicle" gives a summary of everything we know about Nana from both *Nana* and *L'Assommoir:*

> née de quatre ou cinq générations d'ivrognes, le sang gâté par une longue hérédité de misère et de boisson, qui se transformait chez elle en un détraquement de son sexe de femme. Elle avait poussé dans un faubourg, sur le pavé parisien; et, grande, belle, de chair superbe ainsi qu'une plante en plein fumier, elle vengeait les gueux et les abandonnés dont elle était le produit. (215)

Zola uses the character Fauchery as his *porte-parole* in order to explain the contradiction between Nana's beauty and health and the disease and destruction she disseminates: the metaphor of the beautiful plant growing up in a bed of manure serves this purpose quite tidily. The analogy implies, of course, that however beautiful, the flower is always tainted by its roots.

Fauchery goes on to outline the extent of Nana's corruption and powers of destruction:

> Avec elle, la pourriture qu'on laissait fermenter dans le peuple, remontait et pourrissait l'aristocratie. Elle devenait une force de la nature, un

ferment de destruction, sans le vouloir elle-même, corrompant et désorganisant Paris entre ses cuisses de neige, le faisant tourner comme des femmes, chaque mois, font tourner le lait. (215)

It is only at the end of his chronicle, however, that Fauchery introduces the significant and memorable metaphor of the golden fly:

une mouche couleur de soleil, envolée de l'ordure, une mouche qui prenait la mort sur les charognes tolérées le long des chemins, et qui, bourdonnante, dansante, jetant un éclat de pierreries, empoisonnait les hommes rien qu'à se poser sur eux, dans les palais où elle entrait par les fenêtres. (215)

The *mouche* is like Nana in that it is an object of beauty, golden and playful; also like Nana, its golden beauty belies the disease and filth from which it springs and which it communicates to all who are touched by it. The window by which Nana enters polite society is of course the vulnerability created by sexual desire. It is through this weakness that the beautiful but poisoned creature is able to touch men. Perhaps nowhere in the novel is Zola's profound disgust for Nana more clearly revealed than in the analogy between Nana and a disease-ridden insect.

The *fumier* from which Nana comes is both that of poverty and that of the female gender. As both poor and female, Nana (and indeed all prostitutes) represents the ultimate threat to the late-nineteenth-century bourgeoisie.[32] The fear of both physical and moral contamination and disease so prevalent in the period was perhaps the reason for the horrified fascination with prostitution and the idea of the prostitute as the agent of *pourriture; Nana* is without a doubt the most lurid and memorable literary expression of these obsessions.[33]

Perhaps more than any other function it may have in the novel, the "Mouche d'or" narrative serves to provide both the (male) characters in the novel and the readers of the novel with the most fundamental element of the myth of Nana: heredity. Philippe Hamon defines the function of heredity as narrative, and thus as legibility, in Zola:

[Enfin,] l'hérédité est là, implacable, expliquant le conscient comme l'inconscient du personnage; et Zola aime à faire resurgir du passé, en images obsédants ou en souvenir halluciné, les traumatismes qui conditionnent les gestes . . . mettant en lumière la "bête humaine" (Freud ne parle-t-il pas aussi de "la nature animale de notre inconscient"?), cet élément intérieur à l'homme qui n'est pas incommunicable ni impénétrable, mais qui accompagne et explique l'homme, qui fait que sa vie toute entière est "signe" pour une lecture claire, élément dont l'élucidation permet à Zola de récupérer pour son roman les personnages de fous, d'idiots, de névrosés, d'invertis.[34]

One might, of course, add prostitutes to Hamon's list of marginal lives rendered legible by heredity. Nana's life indeed becomes "sign" when read in the context of *L'Assommoir* and of all the preceding volumes of the *Rougon-Macquart* cycle. It is crucial to remember, however, that only we the readers have access to this information; the characters in the novel have only the condensed, allegorized version in the "Mouche d'or." While we are in little danger of failing to recognize and read the system of narrative signs, the unfortunate (male) victims in the novel have but one shot at reading and thereby saving themselves. Zola seems to provide his characters with the explanatory narrative of Nana's evil in the "Mouche d'or" only to demonstrate that, even in the face of such a definitive (if metaphorical) exposition of the danger she represents, they are incapable of reading the account, heeding its warning, and protecting themselves from her.

It is important to note how Zola treats the scene in which the "Mouche d'or" is presented: it is Muffat, the most cruelly exploited of Nana's lovers, who discovers the article while in Nana's boudoir. Perhaps more than any other character in the novel, Muffat is aware of the danger Nana represents; he is also, however, the most stricken with desire for her. Nana, happy for the publicity and seemingly unaware of the horror of the portrait, recommends that he read Fauchery's article. Muffat recognizes his own worst fears outlined in the article and is devastated:

> Muffat leva la tête, les yeux fixes, regardant le feu.
> —Eh bien? demanda Nana.
> Mais il ne répondait pas. Il parut vouloir relire la chronique. Une sensation de froid coulait de son crâne sur ses épaules. Cette chronique était écrite à la diable, avec des cabrioles de phrases, une outrance de mots imprévus et de rapprochements baroques. Cependant, il restait frappé par sa lecture, qui, brusquement, venait d'éveiller en lui tout ce qu'il n'aimait point à remuer depuis quelques mois. (215–16)

As we reach this suspenseful moment in the plot, we have reason to wonder if, in fact, this awakened awareness of Nana's essential nature will reverse (at least for Muffat) the trend of destruction, if the blindness of desire will be cured by knowledge expressed through narrative. The suspense is short-lived, however, and indeed dramatically undone by the very next words of the text:

Alors, il leva les yeux. Nana s'était absorbée dans son ravissement d'elle-même . . .

Muffat la contemplait . . . Le journal était tombé de ses mains. Dans cette minute de vision nette, il se méprisait . . . ne pouvant détourner les yeux, il la regardait fixement, il tâchait de s'emplir du dégoût de sa nudité . . .

Muffat eut un soupir bas et prolongé . . . Brusquement, tout fut emporté en lui, comme par un grand vent. Il prit Nana à bras le corps, dans un élan de brutalité, et la jeta sur le tapis . . .

Il avait conscience de sa défaite, il la savait stupide, ordurière et menteuse, et il la voulait, même empoisonnée. (216–17)

The scene above clearly dramatizes the conflict of the entire novel: the conflict between rational, analytical thought and its antithesis, the confusion and blindness of desire. Muffat, in the process of reading the "truth" about Nana, sees her body and lets the newspaper fall, as he becomes fascinated by the sight of her nudity. As in the rest of the novel, corporality usurps the place of textuality, desire prevents reading.[35] The danger is of course, as we see in this scene, that Nana's body prevents interpretation or, more precisely, renders the power of reason so much weaker than the power of physical desire that it cannot be acted upon.

Nowhere is Muffat's torment more excruciatingly demonstrated than in the scene above, where for at least a moment he has perfect consciousness of the truth about Nana, the monstrosity behind the mask, at the same time as perfect consciousness of the power of his own desire which prevents him from acting on this knowledge. Zola, who uses the "Mouche d'or" as a device for encapsulating and explaining the myth of Nana as femme fatale, shows us that even with hard, written "proof" of Nana's evil in hand, men are overcome by their own desire.[36] The entire process is predetermined and always fatal. The episode plays a key role in the plot by serving as a sort of point of no return.

If the function of the "Mouche d'or" in relation to the rest of the narrative is of considerable significance, so too are the larger implications of its content. The "Mouche d'or," as a narrative of origins, reveals the organizing principles of Zola's deterministic view of human life. This particular worldview is an amalgam of concepts gleaned from Hippolyte Taine, Claude Bernard, and Prosper Lucas, the genetic theorist.[37] What is of greatest interest to me here is the biological and sociological determinism which serves as the basis of Zola's explanation of Nana's essential evil. Zola's theories of heredity (influenced, no doubt, by contemporary, more or less scientific theories, but nonetheless his own) reveal a belief in the inextricability of moral and physical phenomena. His most basic belief, that psychological and moral traits can be transmitted genetically, serves to demonstrate the relation between the two domains.[38] Their relation is never clearly defined, as the two domains seem to alternate in the roles of cause and effect, the moral/psychological creating physiological states (the *détraquement* of Nana's genitals, for example), and the physiological (blood) transmitting moral propensities. Zola never resolves, at least in *Nana*, this "chicken-or-egg" question. The significant gesture is his positing of the physical and the moral as interdependent phenomena, a belief which echoes the concerns of physiognomical thought.

As is always the danger with theories which correlate the physical and the moral, there are political and ethical ramifications to Zola's determinism. To say that Nana is tainted in some kind of essential and predetermined way is to explain her evil by generalizing about her origins and, by extension, the classifications to which she belongs: her family, her social class, her gender. In direct contradiction to Zola's explicit sociopolitical views, which tended to be progressive for the period, the ethos communicated in *Nana* is one of repression, fear, and prejudice.[39] In his overwhelming desire to define human existence as a cause-and-effect system, to ascribe an explanation to human life, Zola falls prey to, and propagates, pernicious myths.

In Nana's case, and indeed in the case of all the Rougon-Macquart, heredity is seen as the genetic transmission of psychosocial disorders. The model is therefore that of contagion, each generation infecting its successor with a propensity to vice in some form or other. Nana's parents were alcoholic and adulterous; her child, while too young to be morally corrupt, is a weak, sickly creature who dies young, clearly the product of "bad blood."[40] Indeed, the hereditary process is reversed and becomes literal contagion when Nana catches smallpox from her own child, who has been more generally "infected" by her from birth. Once again, we find the late-nineteenth-century obsession with communicable disease—this time in the strange guise of defining principle of Zola's myth of heredity. Nana stands as both heir to an inescapable propensity for vice and the agent for propagating such vice; she is but a link in Zola's master chain. However, as a prostitute, as a representative of both the underclass and women, Nana subsumes that which is most staunchly repressed in a repressive society. She is therefore much more than simply one more example of the phenomenon; her myth clearly reveals the most basic ideological tenets of the *Rougon-Macquart* cycle.

It is no coincidence that Nana is the product of the underclass, a class, as we see in *L'Assommoir*, made up of those who have dropped out, or been cast out by circumstance, from the working class. Aside from the logical idea that a woman raised in conditions of economic distress is more likely to be obliged to sell her body than a *bourgeoise*, there seems to be a myth associating economic and social disadvantage with moral depravity. While it may or may not be statistically provable that certain "vices" (including alcoholism and prostitution) occur with more frequency among people with very little money, what most interests me here is the notion, expressed in *Nana*, that a certain class of people is genetically condemned to vice.[41]

To be sure, Zola is well aware of the sociological factors that determine this tendency in certain socio-economic groups. As is the case with much of his worldview, both nature and culture play a role in determining behavior. However, in seeming contradiction with his progressive, reformist sociopolitical views, Zola also seems to believe that at a certain point culture becomes nature and that it is therefore not entirely possible to distinguish the cultural from the genetic. In his own version of evolutionary adaptation theory, Zola seems to believe that several generations of vice breed an inescapable, essential tendency to vice into subsequent generations. Although this is an idea which remains implicit and indeed mysterious in Zola's work, it is very much in keeping with his metaphorical view of society as a body, subject to contagion from disease in any one of its component parts.[42]

The ideological implication of such a view is fairly clear: the disease of vice springs from a particular socio-economic class and is transmitted genetically to those born into that class and by social contagion to the rest of society; the underclass is thus an essentially diseased and dangerous group of people.[43] It must be noted that this idea was a prevalent one in the mid-to-late nineteenth century and frequently appears in discussions of heredity in Zola's time. Bénédict-Auguste Morel provides us with the most succinct expression of these notions in his *Traité des dégénérescences physiques, intellectuelles et morales de l'espèce humaine* . . . (1857):

> Si l'on joint maintenant à [des] mauvaises conditions générales l'influence si profondément démoralisatrice qu'exercent la misère, le défaut d'instruction, le manque de prévoyance, l'abus des boissons alcooliques et les excès vénériens, l'insuffisance de la nourriture, on aura une idée des circonstances complexes qui tendent à modifier d'une manière défavorable les tempéraments de la classe pauvre . . .

Au sein de [notre] société si civilisée, existent de véritables variétés . . .
qui ne possèdent ni l'intelligence du devoir, ni le sentiment de la moralité
des actes, et dont l'esprit n'est susceptible d'être éclairé ou même consolé
par aucune idée de l'ordre religieux. Quelques-unes de ces variétés ont
été désignées à juste titre sous le nom de classes dangereuses.[44]

The passages above suggest a theory of adaptation, in which socio-
economic conditions create corporeal conditions which in turn create
moral states which can, finally, be transmitted genetically. Zola's theories of heredity, as evinced in the "Mouche d'or" scenario, clearly echo
this deterministic line of sociogenetic thought.

Although it may well be true that all socio-economic classes in Zola
are shown to be vice-ridden, *Nana* makes quite clear that it is Nana,
product of the underclass, who is responsible for the unbridled corruption in the novel. If her aristocratic and bourgeois lovers and clients
have propensities to vice of various kinds, it is she who incites them
to act upon them. In *Zola et les mythes*, Jean Borie gives a succinct account of Zola's definition of the underclass as agent of sexuality and
contagion:

On voit se clore le cercle vicieux où se trouve enfermée, non l'univers
romanesque, mais la vision sociale de Zola. Le monde bourgeois, sous le
dérisoire vernis de l'hypocrisie et des manières, est tout entier en proie
au furieux appétit du corps. *Mais le corps par excellence, le corps réduit à lui-
même, c'est, nous l'avons vu, le peuple. C'est donc le peuple qui est à l'origine
de la catastrophe, il est le bouillon de culture à partir duquel la maladie va pouvoir s'étendre:* cette idée est très clairement exprimée dans le diptyque de
L'Assommoir et de *Nana,* par le personnage de la courtisane, sortie des
taudis de la Goutte d'Or et corrompant toutes les classes de la société.[45]

Indeed, one might easily believe that, had Nana's body not driven
them to sexual distraction and financial ruin, most of these men might
well have stayed home by the fireside with wife, children, mental
health, and stable bank accounts intact. Once again, Muffat serves as
paradigm: given his piety and fundamental fear of women, he certainly would have stayed safely in his own *salon;* however, the very
sight of Nana's body leads him on an inevitable course of financial
and moral ruin. It is Nana in fact who indirectly leads to the downfall
of Muffat's wife, Sabine, as well.

The uncanny ability to bring about the ruin and even death of others
through the desire provoked by the mere sight of one's body is of
course the foundation of the femme fatale myth, the myth of the beautiful siren who lures men to their death. *Nana* is but one among many

expressions of the misogynist notion that men are provoked to desire (and, as a frequent consequence, to vice) by women, and therefore it is women who are the original site of sin, and thus they who are at "fault" in matters of sexuality.[46] It is indeed the notion of female sexuality as the locus of corruption, evil, and danger that provides the logic for Zola's narrative, as I shall discuss at greater length below. However, we must note the specific sociopolitical implication of this myth as it is used by Zola as well: *Nana* is the narrative not only of man's corruption by woman, but also to some extent of the corruption of the bourgeoisie by the underclass. By destroying men on both the moral and economic planes, Nana destroys that which defines the late-nineteenth-century bourgeoisie: prudently managed money and rigorously upheld virtue.[47]

While it is undeniably important to recognize Zola's expression of the myth of the genetically determined "classes dangereuses" in *Nana*, it is perhaps even more to the point to recognize the mythology of gender expressed by the novel. *Nana* has been widely read as a sort of misogynist tract, an exposition of pathological fear and revulsion toward women.[48] What is of greatest interest to me here is the myth of woman in *Nana* as a link between the physical and the metaphysical, between a body and an incorporeal, essential being. Once again, as in the preceding chapters, this correlation, a form of physiognomical thought, is put to use as a tool of mythologization of the female gender.

If we read *Nana* in the context of the entire *Rougon-Macquart* narrative, we see that Nana is but one link in a chain of moral and physical depravity. The vice inherent in the family is explicitly defined as being genetically transmitted, and its origin is identified as the ancestress Adélaïde Fouque, known as "Tante Dide," who appears in *La Fortune des Rougon*, the first novel of the *Rougon-Macquart* cycle. Tante Dide is the genetic origin of all familial evil, and it is no coincidence that her sin is that of sexual promiscuity.

Symbol of female sexuality, the Eve of both the Rougon and Macquart families, Tante Dide introduces the famous genetic *fêlure* which serves as the foundation for many of the narratives in the *Rougon-Macquart* cycle, and particularly that of *Nana*.[49] It is perhaps not inappropriate to point out that the very word *fêlure*, used by Zola to describe the genetic propensity to vice in the family, may be read as a vaginal symbol, further reinforcing the idea of female sexuality as origin of all evil. Indeed, even the psychomedical term "hysteria," used to describe the propensity to various social, mental, or moral disorders, is a term which by its very etymology defines the problem as having an origin in female sexuality, in the uterus, to be specific ("hyster-").[50] The *OED* gives the following revealing definition of the word "hysteria":

> 1. *Pathology.* A functional disorder of the nervous system, characterized by such disorders as anaesthesia, hyperaesthesia, convulsions, etc., and usually attended with emotional disturbances and enfeeblement or perversion of the moral and intellectual faculties.
>
> Women being much more liable than men to this disorder, it was originally thought to be due to a disturbance of the uterus and its functions . . .

Even if many of the heirs to Tante Dide's legacy of hysteria are in fact men, the origin of the problem is clearly suggested to be female sexuality. It is worth noting that Nana, the most literal recipient of Tante Dide's sickness, inherits the *fêlure* from her mother, Gervaise Macquart.

In the course of the *Rougon-Macquart* cycle, the *fêlure* takes many forms and is not limited to sexual promiscuity, as is the case for Tante Dide and Nana. The genetic transmission of an essential flaw which may be expressed in various forms and through various behaviors is an idea elaborated on by Morel in his writings on the concept of *dégénérescence*, and was therefore probably a fairly widely known concept in the discussion of heredity in the mid-to-late nineteenth century. Morel's comments echo, and provide an outline of, the concept at work in Zola's theory of heredity:

> Quelle que soit la dégradation physique dans laquelle tombent les buveurs d'alcool et les fumeurs d'opium . . . ce n'est précisément ni ce même cachet de dégradation extérieure, ni les lésions identiques qu'il faudra rechercher dans les descendants . . .
>
> Lorsque nous suivons l'évolution du principe dégénérateur dans les cas où aucune circonstance favorable n'est venue rompre la fatalité qui pèse sur les héritiers d'un mal primitif, nous parcourons une série d'affections nerveuses protéiformes, offrant la plupart du temps un type convulsif, et constituant sous nos yeux ces tempéraments étiolés, souffrants et maladifs, ainsi que ces perversités morales et ces aberrations intellectuelles incroyables qui, par leur fréquence et leur nature, étonnent, à juste titre, ceux qui n'ont pas suivi de près la manière dont se forment les races dégénérées . . . [51]

Morel continues to explain this phenomenon of a sort of nonspecific heredity and its implication for both the physical and moral domains:

Nous n'entendons pas exclusivement par hérédité [de la dégénérescence] la maladie même des parents transmise à l'enfant, dans son développement et avec l'identité des symptômes de l'ordre physique et de l'ordre moral observés chez les ascendants; nous comprenons, sous le mot hérédité, la transmission des dispositions organiques des parents aux enfants. Il n'est pas nécessaire, encore une fois pour démontrer l'existence de cette transmission, que la maladie des parents soit identiquement reproduite chez les enfants: il suffit que ces derniers soient doués *d'une prédisposition organique malheureuse* qui devienne le point de départ de transformations pathologiques dont l'enchaînement et la dépendance réciproque produisent de nouvelles entités maladives, *soit de l'ordre physique, soit de l'ordre moral et parfois des deux ordres réunis.*[52]

Regardless of the questionable scientific veracity of Morel's theory and without delving into actual biographical speculation concerning Zola's readings on the topic, we can see from the above passages that the idea of a genetic, nonspecific weakness was part of the current body of beliefs on heredity in the nineteenth century. Morel's notion of *dégénérescence* parallels Zola's narrative of the consequences of the genetic *fêlure* in the descendants of Tante Dide. Alcoholism, crime, promiscuity, and various emotional, social, and physical disorders, seemingly unrelated, are in fact the results of an original source of corruption, of a *mal primitif.* As we saw above, the original sinner is a woman and her sin that of sexuality, in Zola as in the Judeo-Christian tradition.

Female sexuality as the source of rampant corruption is the very foundation of *Nana.* The correlation of sex, the most basic definition of which is purely anatomical, and an essential metaphysical *mal* is clearly a physiognomical concept. Indeed, Zola suggests in an extremely subtle way in *Nana* that the sexualized (or potentially sexualized) female may display, somewhere on her person, a corporeal sign of her *fêlure.* Muffat's wife, the countess Sabine, first appears as a paragon of wifely bourgeois virtue, but in the course of the novel enters into an adulterous liaison with the journalist Fauchery as well as profligate spending of her husband's fortune. The degradation of a seemingly virtuous woman in a novel such as *Nana* might naturally be read as the suggestion that all women, even the most circumspect, are potentially corrupt. This is in fact almost surely the function of the character of Sabine and the subplot of her fall from grace. Of greater relevance to our purposes, however, is Zola's use of a physiognomical detail to suggest this misogynist ideology.

Nana and Sabine are different physical types, the most apparent difference being the fact that Sabine is dark, while Nana is blond.

However, as Fauchery discovers, they share a curious facial feature: a mole. It is certainly of no small importance that Zola chooses to refer to this dermatological phenomenon with a rather uncommon usage of the word *signe*. It is clearly not too far-fetched to suggest that the *signe* shared by Nana and Sabine corresponds more closely to the first definition of the word given by the *Petit Robert* ("Chose perçue qui permet de conclure à l'existence ou à la vérité . . .") than to any purely dermatological definition.[53] I would like to suggest that the *signe* shared by the two women is an outward sign of female sexuality and depravity.[54]

Obviously, it is a device which must not be read too literally: there are, to be sure, many sexualized females in *Nana* and elsewhere in Zola's œuvre who are not described as having moles. However, the fact that Sabine falls into illicit sexuality demonstrates that she had the potential for such depravity all along, that there was some essential flaw in her from the beginning. The *signe* serves only to effect a rapprochement with Nana, and to "prove" that a woman with the potential for sexual misconduct (and, of course, it can be argued that for Zola all sexual conduct on the part of women is misconduct) will inevitably act on her potential, appearances of decorum notwithstanding. Perhaps Sabine's vice is an innate as Nana's: her father, the elderly marquis de Chouard, is perhaps the most perverse of all Nana's suitors, a geriatric satyr for whom "renifle[r] dans les endroits pas propres" is a veritable passion. The sign, then, may well be a sign of an inherent *dégénérescence*, but it is more specifically female sexuality which links the two women in the novel. Sabine acts as a shadow figure in relation to Nana, and serves to remind us that Nana's vice is not a personal aberration, or even a strictly familial or class-determined pathology; it is the pathology of female sexuality itself.

It is significant that it is Fauchery, the journalist who explains Nana in the "Mouche d'or" article, who recognizes the physiognomical feature shared by the two women:

> Le journaliste . . . resta grave. Mais un signe qu'il aperçut à la joue gauche de la comtesse, près de la bouche, le surprit. Nana avait le même, absolument. C'était drôle. Sur le signe, de petits poils frisaient; seulement, les poils blonds de Nana étaient chez l'autre d'un noir de jais. (89)

Although it is never explicitly stated, the mole serves as a clue for Fauchery in his speculation about the possibility of seducing Sabine. In the end, of course, he does find that the resemblance to Nana was more than skin-deep, and that Sabine is indeed corruptible. The fact that Fauchery notices the mole is contrasted with the fact that Muffat, who has had more opportunity than anyone else to study the two women, seems never to have noticed it at all.

Perhaps the mole was indeed the danger sign that needed to be recognized and read. Perhaps, however, its very obscurity merely suggests that there may exist physiognomical signs, but that they are often too subtle to be read by most "readers," both intra- and extradiegetic. As is the case in Balzac's *La Vieille Fille* (and indeed in Lavater's *Essai*), Zola's novel upholds the notion of the legible body but seems to imply that the act of body reading is virtually impossible for all but the most exceptional would-be readers. Fauchery, a journalist by profession, is the *only* male character in the novel who is capable of recognizing signs, and as such is shown to possess exceptional abilities. It is no coincidence that he is among the very few who escape destruction as well.

Female sexuality is without a doubt the root of all evil in *Nana*. I discussed above the fact that the novel is "about" Nana's body; I would now like to be more specific by suggesting that it is her genitals which serve as the central image in the text. Zola himself states this fact in no uncertain terms in his *ébauche* for *Nana*:

> Devenant une force de la nature, un ferment de destruction, mais cela sans le vouloir, par son sexe seul et par sa puissante odeur de femme, détruisant tout ce qu'elle approche . . . Le *cul* dans sa puissance; le *cul* sur un autel et tous sacrifiant devant. Il faut que ce livre soit le poème du *cul*, et la moralité sera le *cul* faisant tout tourner . . . [55]

Direct references to Nana's genitals and vaginal imagery abound in the text: Nana's vagina, her "petit rien," is the "tool" with which she builds an empire. Jean Borie articulates the specifically genital identity of the "chair centrale" in the text:

> Le caractère, la "psychologie" de Nana est un problème aisément résolu, donc écarté, de façon à laisser le théâtre entièrement disponible pour cet objet . . . Sa personnalité est sans mystère aucun, sans grand charme, et sans prestige, le comble du banal, de l'ordinaire, de l'explicable: tout le mystère est concentré dans l'organe.[56]

Although I would argue that, read as an expression of the hereditary *fêlure*, even Nana's sexual organ holds little mystery, I would agree that it is Nana's vagina which is the agent for the entire intrigue of the narrative.[57] The danger Nana represents to men is a product not of her will, nor of any perverse penchant for cruelty; indeed, she seems almost unaware of the destruction she brings about. The danger seems to come directly from her vagina itself.

The notion of the fatal vagina, expressed throughout the text in a none too subtle fashion (for example, the description of Nana's house as a "grand trou" which seems to swallow everything up, or Muffat, who "disparaît dans la toute-puissance de son sexe"), can of course be read as the fear of castration. In fact, the entire novel might be described as an expression of a morbid fear of castration.[58] What interests me most, however, is the fact that the vagina, the degree zero of female identity from a purely anatomical perspective, is clearly portrayed as the force of destruction. It is the figure of this organ which permits the simplistic, but essential, equation of woman, sexuality, and fatality. When it is said of Nana that "jamais elle n'avait senti si profondément la force de son sexe" (324), the ambiguity of the word *sexe* serves as the clearest illustration of this equation. The logic of *Nana* tells us in no uncertain terms that woman = sex = death.

Nana is a novel which is at once physiognomical and antiphysiognomical. It is physiognomical in that, as we saw above, it is a narrative which has at its center a human body. Zola's beliefs in the genetic transmission of moral, emotional, and social qualities are based on a belief in an essential correlation between the physical and the metaphysical; in this, his entire system of explaining human existence depends on the most basic tenet of physiognomical thought. Furthermore, there are hints of even more explicitly physiognomical thought in the novel, most notably the revelatory *signe* shared by Nana and the countess Sabine. However, the actual function of the *signe* in the text and its implications remain unclear, and we can only speculate that it suggests at least the potential of the body to furnish corporeal signs which reveal essential truths. The practical legibility of such signs is put into question by their subtlety; as we have seen in previous chapters, such is often the case with regard to physiognomical reading.

Unlike Balzac, Zola stops short of advising us to educate ourselves in the art of reading the body. Rather, the central theme of *Nana* is the fact that desire prevents "us" (the male readers to whom Zola's text is undoubtedly addressed) from performing "our" self-defensive duty of reading the (female) body. A crudely carnal enactment of the adage "love is blind," *Nana* shows us that allowing the senses to overpower the intellect is a dangerous, if not fatal, lapse. The novel is perhaps the consummate literary expression of the deeply misogynist myth of the femme fatale, in other words of female sexuality as evil and dangerous. Sexual desire is demonstrated to be antithetical to intellect, and the female body thus the agent of the destruction of (presumably masculine) reason.[59] The characters in the novel, subjugated to blinding sexual desire, completely lack the faculty which would allow them to search for explanation, for a cause-and-effect system of ordering their existence and understanding Nana. Reduced (literally) to following their noses like a "meute de chiens," they abdicate the very faculties which have traditionally been said to separate man from beast. Even when provided with an exposition of the why and how of Nana's evil in the "Mouche d'or" article, they remain prey to their own senses, as Nana's body literally intercedes between them and the printed word.

We as readers have access to Zola's system, both through the "Mouche d'or" and through our (supposed) knowledge of the *Rougon-Macquart* cycle, and can see quite clearly beyond Nana's superficial beauty. Unlike the characters in the novel, then, we possess the logic of narrative, the temporal ordering and explaining of events which renders seemingly disparate phenomena comprehensible. Indeed, Zola creates a system of fictional narrative whose interpretation is built into the text; we are provided with the "answers." In "Zola: romancier de la transparence," Philippe Hamon explains this textual saturation:

> Zola ne multiplie les détails et les preuves que pour mieux conclure, et ne superpose ses effets, n'accumule autour d'une scène ou d'un personnage les explications, les faits corroboratifs réitérés, les histoires exemplaires, les prédictions, les retours en arrière, les *leitmotive*, les développements symboliques ou mythiques, que pour mieux expliciter et dominer à son gré sa matière romanesque.[60]

Hamon goes on to suggest the literary-historical implications of such a predetermination of the interpretation of a text:

> Nous assistons donc chez Zola à un changement important de l'esthétique du Roman: ce n'est plus la multiplicité des sens possibles d'une œuvre donnée qui fait la perfection de cette œuvre, mais la multiplicité des procédés convergents qui y foissonnent et qui concourent à l'élaboration d'un sens simple et non ambigu.[61]

While I am in complete agreement with Hamon's analysis of Zola's system of intratextual explication, we must note in passing that the "elaboration of a simple and nonambiguous meaning" is not Zola's gift to the art of the novel. Nowhere is such a system of authorial domination and predetermination of a single textual interpretation more clearly put into practice than in Balzac. Indeed, this is perhaps the most important of Zola's many debts to Balzac. If Balzac's "pro-cédés" are fewer and more direct than Zola's, his interpretative agenda is the same: the narrative contains its own interpretation, narrative presupposes metanarrative.[62]

Zola, champion of reason and science, of observation and experi-mentation, gives us a novel which, like Balzac's *La Vieille Fille*, serves as a cautionary tale. As we saw above, the literature-as-science agenda of Zola's theoretical text *Le Roman expérimental* does not seem to be prac-ticable, and thus theory and practice are at odds within theory itself.[63] His novel mirrors this contradiction by the fact that it implicitly en-dorses observation and interpretation of the body as physiognomical principles, while rendering them impossible as acts.

Nana thus exemplifies the fundamental rhetorical gesture of the physiognomical novel: its plot clearly indicates that body reading is necessary and possible in theory, while simultaneously—and equally clearly—suggesting that it is impossible in practice for all but the most exceptional "readers" (both intra- and extradiegetic). Nana's depraved disease of hypersexuality is written on her body, implies the text, but the *signe* that signifies it is so obscure as to be virtually impossible to read. Zola's novel points up the paradox of all the physiognomical novels studied here: the truth is indeed inscribed somewhere on the body; reading it is therefore always theoretically possible, but practi-cally improbable on the part of anyone other than the omniscient and impersonal narrator of the text.

Conclusion

I hope that it is fairly clear by now that for all my emphasis on their specificity and seeming heterogeneity, the works I have studied in the preceding chapters share some important ideological and literary foundations. The works were chosen primarily for the thematic centrality of the body they represent; however, as I have explained in the Introduction, these are works in which theme transcends the level of significance traditionally ascribed to it and opens onto larger questions of narrative, representation, signification, and ideology. Each of the works at issue poses some sort of question to which there is an answer; however, the "twist" to these works is that the answer is always to be read on the body of one or another of the characters and that we as readers never have access to the knowledge which would permit us to perform such a corporeal reading.

To borrow Genette's now-standard terminology, these are works whose *histoires* clearly imply physiognomical principles, but whose *récits* offer little to explicate either the principles or the practice of the act of body reading. In this way, then, these works function as allegories of the traditional discourse of physiognomy itself, as exemplified by Lavater. They may thus indeed be called "physiognomical" novels, not simply for the thematics of the legible body they share, but also because as a group they represent an inscription of the rhetorical foundation of physiognomical discourse in fictional narrative.

Physiognomy is consistently defined, both explicitly in treatises on the subject and implicitly (and even, perhaps, unknowingly) in literary works such as these, as a skill which is fundamentally necessary to functioning in the world but whose practices are ultimately inexplicable. Physiognomical thought, it would seem, constitutes an unteachable code. All of the literary texts I have studied (with the exception of Gautier's *Mademoiselle de Maupin*, which has served as a counterexample here) function more or less in the same manner: in order to fully grasp the implications of the work, we must be able to perform what amounts to a physiognomical reading of the fictional body in question. However—and in some cases this is made quite explicit—an active, positive, and informed deciphering of the central body can be performed only by the supremely knowledgeable (and usually unspecified and impersonal) narrator of the text. The reader must necessarily be relegated to a more passive role, obliged to become a spectator of both the story of the body and its interpretation. The only recourse he/she has to what I would consider a more active interpretative role in the entire process is to rise above and analyze these processes themselves. This is what I have attempted to do in *Face Value*.

As I have discussed at some length in the Introduction, the works I have chosen share ideological, as well as rhetorical, agenda. These are novels which implicitly reiterate and indeed extend the kind of misogynist thought found in the physiognomical tradition from pseudo-Aristotle to Lavater. However, it is important to note that rather than the marginalized and secondary object of interpretation and speculation that one finds in physiognomical treatises, women become the central (if invariably problematic) sign to be obsessively scrutinized and deciphered by novelists concerned with the signifying body. The novels studied here treat, either explicitly (*La Vie de Marianne, Mademoiselle de Maupin*) or implicitly (*La Vieille Fille, La Fille aux yeux d'or, Nana*) questions of gender through their exploration of the legible body. One of the things the body is most often purported to signify is gender, a signification of sex that transcends the merely anatomical. Through their narratives of the body, the male authors (with the exception once again of Gautier) frequently imply notions about their female characters, and, by extension, about women in general, that most late-twentieth-century readers probably find offensive. This fact, in and of itself, is neither surprising, illuminating, nor particularly interesting. However, when we realize that all of this ideological content is expressed through the theme of the legible body, it becomes much more worthy of note. If nothing else, the works in question, when read from this perspective, serve as compelling examples of the potential political and social weight of physiognomical thought.

Marivaux's *La Vie de Marianne* is a text whose central question is that of the birth of its eponymous heroine. Our entire understanding of the story must necessarily hinge on whether we believe that Marianne is the child of aristocrats or of their servants. The reader must decide if he/she will accept, as do the central characters of the novel, Marianne's body and its seemingly innate grace and refinement as proof of her aristocratic origins or if he/she will, as do other less central (and less trusting) characters, suspend judgment until more definite proof has been supplied. In making this decision, he/she is in effect handing down a judgment on the notion of physiognomy itself. If we accept Marianne's own description of both her body and the reactions it provokes (and, given the fact that this is a first-person narrative, we are indeed obliged to accept these as givens), we must decide whether or not these primary and secondary corporeal phenomena are indeed reliable signifiers of social class. To some extent, then, our reading of the novel must be predicated on our willingness—or unwillingness, as the case may be—to ascribe powers of signification to a human body.

Significantly, however, the novel remains unfinished, the question of Marianne's birth is never answered, and the reader must therefore continue to wonder about how and what Marianne's body may or may not mean. Our speculation on these questions must remain just that. Marianne the narrator knows the answers, and therefore holds the key to the ultimately "correct" interpretation of her own story, but without a dénouement to the plot, we will never know the truth of her signifying body. The text alludes to, but does not contain, corporeal knowledge, which remains beyond the grasp of its reader.

As well as any other single literary text in this study, Marivaux's *Le Voyageur dans le nouveau monde* can serve as a paradigm of the literary text which turns around physiognomical questions while refusing to offer any discussion or explanation of the concepts they represent. The tale is that of the physiognomical initiation of a young man, who learns that the truth of a person is revealed on his/her body and not through his/her spoken discourse. It is important to note that the tale at first appears to be of the fantasy genre, as the young man assumes that he has traveled to some sort of mysterious parallel universe all of whose inhabitants are perfectly honest. His guide informs him that he has not left the real world but has merely acquired the ability to read people. The acquisition of these physiognomical skills, however, is not explained, either to the reader of the tale or to the character who has miraculously and unknowingly been initiated in the "science." Therefore, physiognomical knowledge is established as being the key to the world, to a higher understanding of human life, but also (curiously) as a skill which is bestowed in some kind of supernatural, inexplicable manner. Thus, the tale remains to some extent linked to the fantastic—the young man's powers of perception are within the realm of human capacity, but it would seem that the principles and practices on which they rest cannot be articulated to the reader of the tale. As in *La Vie de Marianne*, the "answers" to the questions posed by the text are not unknowable, but are nonetheless clearly and unquestionably inaccessible to the reader.

In Balzac's physiognomical novel, *La Vieille Fille*, the question of corporeal knowledge is equally central, if somewhat more subtly presented. As I have argued in Chapter 4, the text very subtly suggests that Mlle. Cormon is something less than a "real woman" and that it is for this reason that she lacks the physiognomical skill required to choose an appropriate husband. She is destined to make the wrong choice and to end up childless. This predestination is, it would seem, legible through the more androgynous features of her body itself (her leg which is like that of a sailor, for example). However, it could hardly be expected that the reader of the novel would know that a certain conformation of the leg, an anatomical part which can not be classified as either a primary or a secondary sexual characteristic, can signify anything about essential gender identity or fecundity. It would seem that the reader can therefore hardly be expected to fully comprehend the story.

The plot of the novel leads us likewise to believe that the chevalier de Valois' masterful nose is a signifier of his great sexual potency, while the text itself remains coy about stating or explaining this particularly central and significant bit of physiognomical wisdom. We must simply assume from the logic of the text that there is indeed a canon of such knowledge to which the text does not provide access. The impersonal narrator, Balzac's amazingly perceptive "observer of the human heart," seems to understand this corporeal logic which directs the events of the story and without which they remain somewhat puzzling. We, however, are not privy to such knowledge. We must simply accept the fact that the answers to the physiognomical questions represented in the text, the keys which will unlock the mysteries of both what Mlle. Cormon is and how she should have made her choice, exist, but are beyond our grasp as mere readers of the novel.

Similarly, in Balzac's *La Fille aux yeux d'or*, there is clearly a connection between the young female protagonist's extremely unusual looks (her golden eyes) and her extremely unusual sociosexual situation (she is a "kept" woman, but kept by another woman; she is a heterosexual virgin, well experienced in lesbian lovemaking). In keeping with the pattern I have been discussing, the nature of this link between the body and the "truth"—sociological, economic, sexual, moral, or other—of an individual character, and the means by which one might gain access to this truth through the body, remain totally obscure. The implication is that one could have known, that one could have read the invisible truth through the visible body, but that to do so would have required special knowledge of a sort not even alluded to within the tale itself. As always, concrete information as to physiognomical practice is unavailable.

In Zola's *Nana*, it is possible for the title character to continue the rampage of moral, intellectual, and physical destruction of which the narrative is composed precisely because the men whom she seduces are blinded by their own carnal desire. As with Balzac's *Vieille Fille*, the misreading (or, more accurately with respect to *Nana*, nonreading) of a body is constitutive of the narrative itself. The logic of the story as a whole—and, indeed, of the entire *Rougon-Macquart* cycle—implies that Nana's corruption and disease are congenital and thus should have been legible somewhere on her body. Such a physiognomical reading would, however, require attention to wildly subtle signifiers which belie her general *air*.

It would appear that the sign of Nana's depravity is, appropriately, her *signe*, a mole. The fact that Nana and her aristocratic counterpart, the initially virtuous but ultimately sexualized and adulterous Sabine de Muffat, have identical moles suggests that they may well be read as signifiers of female (hyper-) sexuality (in Zolian terms, animality and depravity). While such a reading of the novel may at first appear to be far-fetched, I would argue that it is indeed the very obscurity of this all-important *signe* that links Zola's text to the others in this study.

Like Balzac, with whom he shares so many other fundamental concepts, Zola seems to be demonstrating that the body around which his text revolves can indeed be read. However, also like Balzac, such a reading (on the part of either other characters or the reader of the text) would rely on knowledge so obscure as to be impossible to expect from anyone other than a seasoned practitioner of the nonexistent "science" of physiognomy (or any of its various permutations and guises, including Zola's theories of heredity and Balzac's "physiology"). Given the fact that the text itself contains no physiognomical explanation as to how one might interpret Nana's body–for example, a gloss on her famous mole—readers are left clearly outside the "loop" of knowledge the story clearly, if subtly, represents. Physiognomical signs, it would seem, are definitely present but definitely beyond our grasp. Nana's mole functions as a doubly significant sign, representing not only her depraved sexuality, but also, through its very obscurity, the impossibility of performing the kind of corporeal reading that would have had to be undertaken in order to recognize her innate evil. The mole indeed serves as a sort of interpretative "black hole," suggesting at once signification and its impossibility.

As should be obvious from both my Introduction and Chapter 5, Gautier's *Mademoiselle de Maupin* is intended for the purposes of this study to serve as a clear counterexample to the phenomena I have outlined above. It is indeed, I would maintain, an antiphysiognomical novel, in that it revolves around questions of body reading, ultimately to endorse only a purely subjective, amoral, esthetic response to human bodies. Specifically, it refuses any received notions about the relation between sex and sexuality, by concluding with a coyly affirmative response to the question of same-sex desire it has posed from the start. Such conclusions could hardly be farther from the kind of simultaneously scientistic and moralistic agenda alluded to, and enacted by, Balzac's and Zola's texts. *Mademoiselle de Maupin* clearly puts into question the notion that bodies can be made to signify, and therefore refuses to posit certain corporeal knowledge of any kind. The reader's relationship to such a "modern" text is entirely different from the one I have described at work in the texts above, in which some superior knowledge is held over the head of the reader, rendering him/her passive and confused with respect to the role of the signifying body in the text. In sharp contrast to the other works in this study, Gautier's novel informs us not of what we do not know, but rather of what cannot be known.

Ultimately, then, the physiognomical novels I have chosen to study here are linked not only by "mere" physiognomical theme but, more important, by the dynamics of reading, rhetoric, and representation they suggest. As I have attempted to demonstrate with respect to Lavater, all physiognomy per se is based on the assumption of a field of knowledge which is curiously never articulated or elucidated. Lavater purports to teach his readers the principles and practices that constitute the would-be science, but fails to do so. The student of physiognomy is thus left with the curious impression that physiognomical readings can be performed but not explained. The reader of Lavater's "fragments" can be a witness to the author's interpretations of the many representations of the body the work contains, but is excluded from participating in this particular interpretative process him/herself. The reader therefore cannot *become* a physiognomist, even (especially) by studying Lavater's master text.

La Vie de Marianne, La Vieille Fille, La Fille aux yeux d'or, and *Nana* are novels which are driven by an analogous dynamic concerning knowledge, the text, and the reader; indeed, they represent the inscription of the fundamental rhetorical gesture germane to all physiognomical discourse into the realm of fictional narrative. An imperfect but provocative way of characterizing this rhetorical gesture as inscribed into fictional narrative would be to suggest that these are works which represent a notable breakdown in the relation between *récit* and *histoire.* On the level of *histoire,* of plot, physiognomical reading is central, while on the level of *récit,* of narrative, it remains obscure. The reader of these novels cannot but be aware of the importance of body reading and the codes required to perform it; however, the text in question itself offers no exposition or explication of these codes.[1] Traditional "realist" narrative works are grounded in a mutual and reciprocal relation between *récit* and *histoire:* the text not only presents but explains the story, and the story in turn informs our understanding of the text. The novels here represent a failure in this chain; the text obviously presents the story, but offers little of the interpretative or didactic information without which the story remains at least partially incomprehensible.

In keeping with the physiognomical tradition, then—and no doubt in most cases completely unbeknownst to the authors themselves— each is a text which foregrounds the human body, poses some very pointed questions about its potential legibility, all the while ensuring that its reader will lack the knowledge necessary to respond to those questions. Reflecting the most basic agenda of more or less "realist" fictional narrative, on the other hand, these novels purport to explain the world and human existence. They take as their primary tool in this enterprise the human body. It is therefore particularly ironic that, rather than works of explication, these physiognomical novels turn out to be works of obfuscation.

As I have discussed in Chapter 4 in relation to Balzac, the German theorist of literary reception Wolfgang Iser has argued that the literary (especially narrative) work exists only through the response of its reader to certain "indeterminacies" it contains.[2] According to such theories, the act of reading itself consists of the resolution on the part of the reader of these indeterminacies. The texts I study represent a strange subgenre of narrative texts in which certain thematic "indeterminacies" are preemptively resolved by the texts themselves, thereby preventing the reader from performing the kind of interpretative acts which Iser considers to constitute reading itself. The interpretative processes through which such interpretations may be arrived at are themselves obscured, thus rendering these texts doubly "unreadable."

It bears repeating that the inescapable irony is that these are works which thematize, in some cases obsessively, the acts of deciphering, reading, and interpreting. Their literary strategies are thus in direct contradiction to their thematic concerns. It is significant that it was my interest in these works as expressions of a certain set of themes that first drew me to them. I did not know at the outset that these thematic similarities would be mirrored by the more fundamental, more complex similarities I have suggested both here and in the Introduction. Indeed, as it happens, these particular raw themes serve as allegories of the texts themselves as legible (or illegible) objects. I would like to think that my work may therefore be seen as an object lesson in the study of the relation of theme and function in works of narrative literature. I would speculate that the thematic plane can indeed often be coaxed to open onto literary questions whose interest transcends that of theme itself. This would suggest that an explicit rehabilitation of theme as an operative principle in literary analysis should be considered.

I hope the preceding chapters have demonstrated in a persuasive manner that thematic readings can serve as foundations for complex rhetorical and narratological analysis, particularly when the themes in question are those of reading, interpretation of signs, intelligibility, and knowledge. A theme such as that of physiognomy is clearly of much more interest as a figure of rhetorical and ideological interpretation than merely as an offbeat footnote to cultural history. Physiognomical thought, particularly when defined as broadly as I have here, presents vast avenues of potential exploration, especially for scholars of eighteenth- and nineteenth-century literature. My goal in writing *Face Value* was to facilitate, rather than preclude, future treatments of this compelling and important question.

Ultimately, I would like to suggest the supreme irony underlying the innumerable failed attempts to articulate physiognomical thought: that the human body, the most familiar object of immanent experience, should prove to be the most elusive object of interpretation.

Notes
Bibliography
Index

Notes

Introduction

1. It is worth noting that much poststructuralist criticism already uses theme as jumping-off place (the themes of reading and writing, the book, the letter, etc.) for "allegorical" readings. Interestingly, however, the role of theme itself has rarely been explicitly discussed in such analyses.

2. This definition of the basic function of narrative discourse is adapted from that of Peter Brooks, specifically from his *Reading for the Plot: Design and Intention in Narrative* (New York: Vintage Books, 1985). I will follow Brooks's general principles while molding them of course to my own ends.

3. Helena Michie identifies a similar dynamic in certain Victorian novels, whose questions are more pointedly sociopolitical than those of most of the texts I discuss here (with the notable exception of Zola, whose lower-class femme fatale plot mirrors those Michie discusses in British literature of the same period): "The heroine's body figures prominently in a series of Victorian texts that present a social or criminal problem; the body itself becomes a series of clues to solving a central mystery, the correct reading of the body an answer to pivotal social questions." See *The Flesh Made Word: Female Figures and Women's Bodies* (New York and Oxford: Oxford University Press, 1987), 119.

4. See Peter Brooks, *The Melodramatic Imagination: Balzac, Henry James, Melodrama and the Mode of Excess* (New Haven: Yale University Press, 1976).

5. See Michie, *Flesh Made Word*, 7: "Women's bodies occupy no . . . stable and comforting relation to the unknown; since they are themselves the unknowable, the unpenetrable mystery, they are not so much vehicles of epistemological consolation as they are sources of change, disruption, and complication." I will have occasion to cite this same passage from Michie again below, when I discuss women as objects of physiognomical thought in my chapter on the history of the would-be science.

6. Barbara Maria Stafford, *Body Criticism: Imaging the Unseen in Enlightenment Art and Medicine* (Cambridge, MA, and London: MIT Press, 1991), 126. I would point out that this "male" desire for order and intelligibility as expressed through physiognomical thought is not exclusive to the classical and neoclassical periods of intellectual history. Oddly, it seems to be a constant of human thought across the centuries. It is worthy of note that, to my knowledge at least, there is no record of any female physiognomists throughout the many centuries in which various forms of the would-be discipline have been practiced.

7. See Jean-Jacques Courtine and Claudine Haroche, *Histoire du visage: XVIe–XIXe siècle* (Paris: Rivages, 1988); Philippe Perrot, *Le Corps féminin: le*

travail des apparences XVIIe–XIXe siècle (Paris: Fayard, 1984); Daniel Roche, *La Culture des apparences: une histoire du vêtement: XVIIe–XVIIIe siècle* (Paris: Fayard, 1989); and Stafford, *Body Criticism.*

8. Stafford, *Body Criticism,* 84.

9. Roche, *La Culture des apparences,* 388.

10. Graeme Tytler, *Physiognomy in the European Novel: Faces and Fortunes* (Princeton: Princeton University Press, 1982).

11. Roger Kempf, *Sur le Corps romanesque* (Paris: Seuil, 1968), 7.

Chapter One. From Analogy to Causality: The History of Physiognomy before 1700

1. I was greatly informed by, and am thus greatly indebted to, Jean Bottéro's essay on physiognomy, medicine, and scientific method in Mesopotamia for its wealth of information and insight concerning these earliest avatars of physiognomy. See Jean Bottéro, "Symptômes, signes, écritures," in *Divination et rationalité* (Paris: Editions du Seuil, 1974). Bottéro frequently acknowledges F. R. Kraus, *Die physiognomischen Omina der Babylonier* (Leipzig, 1935) as his source for information concerning physiognomical documents of the period. Large fragments of a large Mesopotamian treatise on physiognomy (in the form of tablets) still exist.

2. Bottéro, *Divination et rationalité,* 107, rightly distinguishes between the two as "physiognomonie au sens propre" (study of the body) and "physiognomonie au sens large" (study of behavior).

3. Ibid., 191–92.

4. Ibid., 191.

5. Ibid., 86.

6. It is, however, to be found in Aristotle, *Complete Works,* ed. Jonathan Barnes (Princeton: Princeton University Press, 1984), 1: 1237–50.

7. It is interesting to note here that the first known system of classification of animals is Aristotle's, from the fourth century B.C.

8. Helena Michie characterizes the particular place women have occupied in the history of the signifying body: "Women's bodies occupy no . . . stable and comforting relation to the unknown; since they are themselves the unknowable, the unpenetrable mystery, they are not so much vehicles of epistemological consolation as they are sources of change, disruption, and complication." See *The Flesh Made Word,* 7.

9. It is for this method that the pseudo-Aristotelian treatise is most often remembered.

10. It should be specified that the author of the treatise himself issues a disclaimer for the final method: "The last mentioned method [that of pathognomy] by itself, however, is defective in more than one respect. For one thing, the same facial expression may belong to different characters . . . Besides, a man may at times wear an expression which is not normally his . . . and, thirdly, the number of inferences that can be drawn from facial expression

alone is small" (1237–38). This dismissal of pathognomical observation as opposed to physiognomical observation will be germane to Lavater's method as well.

11. For a survey of physiognomical and other related writings in the period between pseudo-Aristotle and Lavater and their possible influence on Lavater, see Tytler, *Physiognomy in the European Novel: Faces and Fortunes*, 38–54. For information on changing usages of words referring to the face and the changing mentalities they reflect, see Jean Renson, *Les Dénominations du visage en français et dans les autres langues romanes* (Paris: Les Belles Lettres, 1962), 164, 188.

12. See Tytler, *Physiognomy in the European Novel*, 36.

13. See Jacques André's introduction to the anonymous Latin *Traité de physiognomonie* (Paris: Les Belles Lettres, 1981), for a good, brief survey of physiognomy in Greece and Rome.

14. See Tytler, *Physiognomy in the European Novel*, 36.

15. The "humors," as described in medical texts as early as those of Hippocrates (c. 450–c. 370 B.C.), were the four basic fluids thought to be found in the human body: blood, phlegm, bile, and black bile. From Plato (427–347 B.C.), each one was also thought to determine a corresponding personality type in those whose bodies it dominated. In a corollary to the theory, "temperament" was the equilibrium of the humors in an individual; disease was thought to be caused by their disequilibrium. As a theory which often attempts to articulate the relation between physiology and personality (or, as it was more frequently described, between the body and the soul), the notion of the humors has a link to physiognomical thought which is fairly obvious. The theory of humors is considered to have been the basic physiological theory from Antiquity to the 18th century. Interestingly, modern developments in medicine (particularly in biochemistry and the study of hormones) have not completely repudiated the basis of the ancient theory. For succinct and informative definitions of these terms, see *Dictionary of the History of Science*, ed. W. F. Bynum, E. J. Browne, Roy Porter (Princeton: Princeton University Press, 1981), 191–92 and 417.

16. See André, introduction to *Traité*, 11.

17. See ibid., 12–13, translating from Cicero: "elle [la nature] a tracé de telle sorte les lignes de sa physionomie qu'elle y a empreint les tendances secrètes de son caractère." (This statement comes in a passage in which Cicero is discussing the nature of man in general.) Also: "C'est l'âme, en effet, qui anime toute l'action, et le miroir de l'âme, c'est la physionomie, comme son truchement, ce sont les yeux."

18. Ibid., 15.

19. Ibid., 17. Physiognomy, as men writing to men, might indeed be called a "homosocial" discourse, to borrow Eve Sedgwick's term. See Eve Sedgwick, *Between Men: English Literature and Homosocial Desire* (New York: Columbia University Press, 1985).

20. André, introduction to *Traité*, 17–21 and 51–56.

21. As I shall attempt to point out in later chapters, physiognomical thought and murky, essentialistic definitions of gender seem to be inevitably linked.

See Judith Butler, *Gender Trouble* (New York: Routledge, 1990), for an important and original critical analysis of the distinction between the categories of "sex" and "gender."

22. André, introduction to *Traité*, 18.

23. See Courtine and Haroche, *Histoire du visage*, 54–55.

24. See ibid., 67.

25. See Tytler, *Physiognomy in the European Novel*, 41.

26. Giovanni Batista della Porta, *Della fisonomia dell'huomo* (Venice: Presso C. Tomasini, 1644), 1.

27. Courtine and Haroche, *Histoire du visage*, 69.

28. Although Porta has long been viewed by scholars as influential and innovative, Louis Van Delft convincingly argues that Porta's significance has been greatly exaggerated. See Van Delft, "Physiognomonie et peinture du caractère: G. Della Porta, Le Brun, La Rochefoucauld," *L'Esprit créateur* 25, 1 (Spring 1985): 43–52, esp. 43–45.

29. René Descartes, *Les Passions de l'âme*, ed. Geneviève Rodis-Lewis (Paris: Vrin, 1966), 63.

30. Expression will later be separated from morphology by Lavater with the introduction of a distinction between physiognomy (fixed features) and pathognomy (mobile features). In the nineteenth century, of course, Darwin picks up the study of both expression and human/animal comparisons.

31. One of the interesting places where the text shows up is in the Moreau de la Sarthe edition of Lavater (1820).

32. Charles Le Brun, "Conférence sur l'expression des passions," *Nouvelle Revue de psychanalyse* 21 (Spring 1980): 95–96. See also the article on Le Brun by Hubert Damisch which follows the *conférence*, "L'Alphabet des masques," 123–31.

33. On Le Brun's relation to Descartes (specifically to *Les Passions de l'âme*), see Van Delft, "Physiognomonie et peinture du caractère," esp. 48–49. Van Delft cites an unpublished thesis from the University of London, J. Montagu's "Charles Le Brun's 'Conférence sur l'expression générale et particulière'" (1959), as the source for much of his information concerning Descartes's influence on Le Brun. Van Delft also cites H. Souchon, "Descartes et Le Brun: Etude comparée de la notion cartésienne des 'signes extérieurs' et de la théorie de l'Expression de Charles Le Brun," *Les Etudes philosophiques* (Oct.–Dec. 1980). See also Stephanie Ross, "Painting the Passions: Charles Le Brun's *Conférence sur l'expression*," *Journal of the History of Ideas* 45, 1 (Jan.–March 1984): 25–47, esp. 27–32. These articles obviously articulate at much greater length, and thus with more subtlety, the relation between the two works than do I.

34. See Ross on the differences between Descartes's and Le Brun's definitions of some of the passions. I, along with most critics, find the similiarities to be much more marked than the differences. Ross nonetheless provides an invaluable account of the different ways of evaluating Le Brun's enterprise. The definitions of "tristesse" and its physiology are to be found in Descartes, *Passions*, 132, 138, and, reproduced by Le Brun in his "Conférence," 97, 98.

35. Curiously, this passage is part of an English translation of Le Brun's

conférence, A Method to learn to design the Passions, proposed in a conference on their General and Particular Expression, trans. John Williams, esq. (London, 1734), but not in the version reproduced by Damisch in *Nouvelle Revue de psychanalyse*, hence the English quotation.

36. As Bottéro tells us, in describing the "classic" Mesopotamian treatise: "Lorsque l'objet en question ou l'un de ses aspects étaient à la fois difficiles à décrire et faciles à représenter, il arrivait qu'un croquis, dans le texte, en précisât la forme en vue" (*Divination et rationalité*, 85).

37. The titles of some of the English translations of Le Brun alone demonstrate this fairly well: *Heads representing the Various Passions of the Soul; as they are expressed in the Human Countenance: Drawn by that Great Master Mons. Le Brun and finely Engraved on Twenty Folio Copper Plates; nearly the Size of Life* (London: Laurie & Whittle, 1794); *A Series of Lithographic Drawings Illustrative of the Relation between the Human Physiognomy and that of the Brute Creation: From Designs by Charles Le Brun, with Remarks on the System* (London, 1827). Le Brun's drawings of facial expression are also well-known for having been used as illustrations for the article "Dessin" in the *Encyclopédie*. I would like to thank Charles Porter for having pointed this out to me.

38. See Ross, "Painting the Passions," 29–30.

39. Interestingly, pseudo-Aristotle himself ascribed great importance to both movement and facial expression, in contradiction to his belief that the fixed features were more reliable than the mobile. See Aristotle, *Complete Works*, 1239.

40. See Courtine and Haroche, *Histoire du visage*, 94–99. Damisch calls the head in repose (that of "tranquility") the *degré zéro* of the expression of passion in Le Brun. See "L'Alphabet des masques," 123.

41. I am greatly indebted to Courtine and Haroche's insightful, sophisticated (if abstract) analysis of the shift in physiognomical thought represented by Descartes and Le Brun. See *Histoire du visage*, esp. 89–102.

42. Ibid., 98–99.

43. Marin Cureau de la Chambre, *Les Charactères de passions* (Paris: Chez P. Ricolet, 1653), Avis (no page number).

44. Indeed, the fifth of Cureau's five "règles générales" is that of physiognomical syllogism.

45. Marin Cureau de la Chambre, *L'Art de connoistre les hommes . . .* (Paris: Chez Jacques d'Allin, 1667), 47.

46. Ibid., 50–51.

47. Peter Brooks, *The Novel of Worldliness* (Princeton: Princeton University Press, 1969). For a convincing analysis of the growing exploitation of physiognomy for social purposes, see Courtine and Haroche's reading of Cureau's *L'Art de connoistre les hommes . . .* (*Histoire du visage*, 31–34).

Chapter Two. Marivaux, *le masque*, and *le miroir*

1. For a survey and discussion of the precursors (from the Renaissance through the eighteenth century, including the *précieux* and the moralists) of

such would-be sociology, see Peter Brooks's chapter "The Proper Study of Mankind" in his *The Novel of Worldliness,* 44–93.

2. In his article "Les Machines de l'opéra: le jeu du signe dans *Le Spectateur français* de Marivaux," *French Studies* 362 (1982): 154–70, G. P. Bennington says that the primary work of the *Spectateur* is to demystify by separating signs (gestures, for example) into their component parts. Bennington also makes the very valid point that the work does not take into account the fact that, as writing, it is itself subject to all the same fallacies inherent to the act of signifying.

3. This focus on man-in-society (particularly in the codified, closed segment of society known as "le monde") in certain works of eighteenth- and nineteenth-century literature is the object of Brooks's study cited above. These works, in effect, constitute a subgenre that Brooks aptly calls "the novel of worldliness."

4. The ubiquitous metaphor of the mask, using the true face as metaphor for the true character, is obviously one born of physiognomical thinking.

5. Ronald Rosbottom, *Marivaux's Novels: Theme and Function in Early Eighteenth-Century Narrative* (Rutherford, N.J.: Fairleigh Dickinson University Press, 1974). Rosbottom's was the first book-length work of in-depth literary criticism to focus exclusively on Marivaux's narrative works.

6. See Henri Coulet, *Marivaux romancier* (Paris: Armand Colin, 1975), 503–4, for an inventory of quotations from seventeenth- and eighteenth-century novels showing *coquettes* before their mirrors. See Philip Stewart, *Le Masque et la parole: le langage de l'amour au XVIIIe siècle* (Paris: José Corti, 1973), 129–30, who notes the symbolic importance of this scene. Stewart suggests that almost all of Marivaux's literary work is founded on the "ambivalence" created by this scene.

7. Rosbottom, *Marivaux's Novels,* 80.

8. Frédéric Deloffre and Michel Gilot's edition of Marivaux's *Journaux et œuvres diverses* (Paris: Garnier, 1969), 118; emphasis mine. All subsequent page references are to this edition.

9. He is also perhaps using the convenient device of speaking through a mouthpiece (the "indigent philosophe" is the most cynical and radical of all the periodical narrators) to pronounce views somewhat more extreme than his own probably were, and certainly, as I have stated, in contradiction with those implicit in *Marianne.*

10. The notion of the existence of a language less arbitrary and therefore more reliable than that of words is one which Lavater will espouse and which is fundamental to physiognomical thought in general. It is discussed at some length in the following chapter on Lavater.

11. In his "Marivaux in the Classical Episteme, or: Suis-je chez moi, Marie?" *L'Esprit créateur* 25, 3 (Fall 1985): 30–41, James Creech gives a completely different interpretation of the tale. Using the Foucaldian definition of the classical *episteme* as an antimodel, Creech argues that *Le Voyageur* refutes, rather than affirms, the transparency of the sign. See also Georges Van Den Abbeele, "La Vérité en négligé: le 'Nouveau Monde' de Marivaux," *French Studies* 40,

1 (1986): 287–303. For his part, Philip Stewart points out that the insincerity and requisite unmasking learned by the traveler-initiate in *Le Voyageur* are diversions, more amusing than tragic. See Stewart, *Le Masque et la parole*, 124–25.

12. In "Le Voyageur dans le Nouveau Monde," the narrator-initiate is indeed given books from which to study human character. However, when he actually acquires the ability to divine character, it is such an instinctive act that he does not even realize that he is performing it.

13. Suzanne Muhlemann, *Ombres et lumières dans l'œuvre de Pierre Carlet de Chamblain de Marivaux* (Bern: Editions Herbert Lang, 1970), 55.

14. Henri Coulet notes, in *Marivaux romancier*, that usually Marivaux's characters say that something is difficult or impossible to express in words just after they have done so quite effectively (301–2).

15. Marivaux, *La Vie de Marianne*, ed. Marcel Arland (Paris: Garnier-Flammarion, 1978), 49–50; all subsequent quotes from this edition unless otherwise indicated.

16. See Brooks, *Novel of Worldliness*, 132, for the idea that "*sentiment* and *cœur* are usually equivalents for intuition in Marivaux . . ."

17. Marianne's skill as a spectator has been commented on by many critics. David Coward, in his *Marivaux: La Vie de Marianne and Le Paysan parvenu* (London: Grant and Cutler, 1982), says: "She [Marianne] looks long and hard at those whom she encounters, interpreting their gestures and facial expressions, and translating what they say into what they really mean . . ." (21); "For . . . Marianne, human beings are transparent and, fifty years before Lavater tried to justify the reading of character from the interpretation of physiognomies, Marivaux tests to the limit our belief that our true natures are written on our faces . . . With deadly efficiency, they [Marianne and Jacob in *Le Paysan parvenu*] strip their victims of their cultivated manner and social personalities. They are endowed with an instinctive lucidity compounded of affective intuitions and a few simple moral standards . . ." (25–26). In *Le Thème de l'être et du paraître dans l'œuvre de Marivaux* (Zurich: J. Druck, 1969), Harold Schaad says that "le regard interprétatif va atteindre un degré de subtilité tout à fait étonnant. Il deviendra l'essentiel du personnage marivaudien . . . on échange des regards chargés de signification" (22).

18. For a clear and succinct analysis of the relation between the seventeenth-century *portrait* (the *précieux*, the *moralistes*) and the *portrait* as it appears as set piece in the eighteenth-century novel (particularly those belonging to the subgenre he calls "the novel of worldliness," including *La Vie de Marianne*), see Brooks, *Novel of Worldliness*, esp. 48–68, 103–15.

19. These references to Hamon's work are very aptly cited by Hendrik Kars in *Le Portrait chez Marivaux: étude d'un type de segment textuel* (Amsterdam: Rodopi, 1981), 123. The essays in question by Hamon are: "Clausules," *Poétique* 24 (1975): 495–526, and "Texte littéraire et métalangage," *Poétique* 31 (1977): 261–84.

20. See Coulet, *Marivaux romancier*, 323.

21. *Marivaudage*, often used pejoratively to refer to the overly refined

psychological analysis in Marivaux's œuvre, is defined by the *Petit Robert* (1984) as follows: "Affectation, afféterie, préciosité, recherche dans le langage et le style (attibués à Marivaux)." Although this stylistic phenomenon has been commented on by many critics, the definitive work on the subject is Frédéric Deloffre, *Marivaux et le marivaudage* (Paris: Armand Colin, 1967).

22. Brooks comments on the intertwining of the physical and the moral in this portrait (*Novel of Worldliness*, 108–09), and remarks that the physical and the moral are finally shown in light of their social results (109). For his close readings of the portraits of Mme. de Miran and Mme. Dorsin, see 103–15.

23. It is important to remember that, for the purposes of this study, I am considering pathognomy to be a subset of physiognomy, and not a completely different category of study as Lavater does.

24. See Coulet, *Marivaux romancier*, on the terms *air* and *physionomie* in these portraits and how his usage prefigures Lavater: "La physionomie est moins souvent mentionnée que l'air . . . mais elle figure toujours dans les portraits dont elle est un élément; ailleurs, si elle ne se confond pas tout à fait avec l'air, l'air semble être une modification de la physionomie, ou un aspect passager qui le contredit ou le confirme. Le mot lui-même semble assez évocateur pour Marivaux, il signifie que le visage est particulièrement expressif . . . Marivaux a une idée plus distincte de cette physionomie qu'il nommait déjà dans ses premiers romans . . . il songe à quelque œuvre où 'ce terme de physionomie' est un terme technique" (312).

25. Although there are fundamental distinctions to be made between the moral and social domains, particularly in the context of the portrait, what interests me here is the more general distinction between the physical self and the metaphysical (or nonphysical) self, which includes both the moral and social selves.

26. I obviously use the term "metaphysical," here and elsewhere in this study, in its etymological sense (with a small *m*).

27. Kars, *Le Portrait chez Marivaux*, 29.

28. Coulet, *Marivaux romancier*, 318. Coulet also makes the very valid point that the portraits depict the atemporal, static essence of a character, and not his evolution (317). This observation provides a further link between the portraits and physiognomy, that of a concern with essence as opposed to personality. See also Kars, *Le Portrait chez Marivaux*, 100.

29. Vivienne Mylne, *The Eighteenth-Century French Novel: Techniques of Illusion* (Manchester: Manchester University Press, 1965), 118.

30. Kars, *Le Portrait chez Marivaux*, 100. For an informative and different view which distinguishes between the various types of narrative stasis in *Marianne*, see Philip Stewart, "Marianne's Snapshot Album: Instances of Dramatic Stasis in Narrative," *Modern Language Studies* 15, 4 (Fall 1985): 281–88.

31. I hope my readers will excuse my frequent use of the rather unwieldy "him/her" construction to refer to the reader here. In the context of a discussion of the narrative strategies of a text whose *narrataire* is a woman, the pseudo-general "he" seems particularly inappropriate.

32. Marcel Arland, *Marivaux* (Paris: Gallimard, 1950), 55.

33. Jean Molino, "Orgeuil et sympathie à propos de Marivaux," *Dix-Huitième Siècle* 14 (1982): 348.

34. There are suggestive comparisons and contrasts to be made between the treatments of the themes of physiognomy and physical appearance in *La Vie de Marianne* and *Le Paysan parvenu*. Both Jacob and Marianne survive in the world on their good looks and charm; the facts of Jacob's lowly birth are, however, never put in question and his physical charm must therefore stand on its own merit. An analysis of the relation between the two novels might well take the difference in gender of the two title characters as its basis. See, among others, Marie-Anne Arnaud, "*La Vie de Marianne* et *Le Paysan parvenu*: itinéraire féminin, itinéraire masculin à travers Paris," *Revue d'histoire littéraire de la France* 82, 3 (1982): 392–411; Frédéric Deloffre, "De Marianne à Jacob: les deux sexes du roman chez Marivaux," *L'Information littéraire* 11, 5 (Nov.–Dec. 1959): 185–92.

35. In *The Surprising Effects of Sympathy* (Chicago: University of Chicago Press, 1988), David Marshall effectively interprets the prevalence of themes of sympathy, pitiful spectacles, and theatrical relations in Marivaux's non-dramatic work (including *Marianne*) and elsewhere in French and English literature. He analyzes this scene of encounter between Marianne and Mme. de Miran as a spectacle, with Marianne the (sympathy-evoking) spectacle and Mme. de Miran the spectator (70–72). At one point, Marshall makes the astute and suggestive speculation that the name "Miran" "shares with *miroir* an etymology in *mirer*: 'regarder avec attention' . . ." (71). One can also easily imagine a reading of such dramatic scenes in Marivaux's novels juxtaposed with specific analyses of his comedies.

36. The theme of the orphan or *enfant trouvé* was something of thematic convention, particularly in the *roman héroïque*. See Coulet, *Marivaux romancier*, 378, and Coulet and Michel Gilot, *Marivaux: un humanisme expérimental* (Paris: Larousse, 1973), 172–75.

37. One might speculate that Mme. de Miran's sympathy is born of an identification with the young girl which would be unlikely if she did not assume the girl to be of her own social class. Indeed, Philip Stewart asserts that Mme. de Miran's sympathy is "predicated on like birth . . ." (*Half-Told Tales: Dilemmas of Meaning in Three French Novels* [Chapel Hill, 1987], 28).

38. The idea that the blood transmits certain physical, intellectual, and social traits is of course an accepted commonplace (indeed, the organizing principle) of any aristocratic society. See Marshall, *Surprising Effects of Sympathy*, 50–51, for a very suggestive double reading of Marianne's phrase "le sang d'où je sortais" which includes both the figurative meaning of *sang* (as in heredity) and the literal blood from which the young Marianne was extracted at the time of the fateful coach incident. In *Half-Told Tales*, Philip Stewart reads this same passage and astutely remarks that, in the context of such a discourse as Mme. de Miran's, blood "can only be of one kind" (noble) (22).

39. Philip Stewart has articulated the blurring in *Marianne* of the two traditionally distinct categories: "One really cannot know whether merit or birth is principally at issue: the first seems to gain ground on the second as the

story progresses, but ultimately the second comes to corroborate the first."
See *Half-Told Tales*, 24–25.

40. Stewart makes the very useful distinction between Marianne the narrator and Marianne the character by calling the character "Marianne I" and
the narrator "Marianne II." He notes that the "nobility" of Marianne I makes
plausible the fact that, as we know, Marianne II is "one way or another, noble."
See *Half-Told Tales*, 26.

41. Arland, *Marivaux*, 60.

42. Coulet quotes this passage by Arland and comments that "La dernière
proposition n'est pas tout à fait sûr, surtout si l'on considère en Marianne non
pas l'art de parvenir, mais la délicatesse de cœur et de l'esprit . . ." (*Marivaux
romancier*, 224).

43. Stewart analyzes the centrality of essentialism in *Marianne*, and the inextricability of the notions of origins and essences which provide the basis of
the novel. See *Half-Told Tales*, 30–31. This essentialism is obviously in direct
contradiction with many of Marivaux's writings in which he espouses more
or less democratic ideals, among them "L'Education d'un prince" (*Journaux et
œuvres diverses*, 513–28) and, indeed, *Le Paysan parvenu*.

44. For comments on the role of feminine beauty in the rococo esthetic, the
"centrality of the feminine principle in the rococo aesthetic," and the relation
of the themes of *Marianne* and the rococo, see Patrick Brady, "The Concept
of Rococo Style in Literature," in *Structuralist Perspectives in Criticism of Fiction: Essays on Manon Lescaut and La Vie de Marianne* (Bern: Peter Lang, 1978),
85–114.

45. Leo Spitzer, "A Propos de *La Vie de Marianne*, lettre à M. Georges
Poulet," *Romanic Review* 44 (1953): 121. Stewart, while recognizing that Marivaux's is a "stereotyped discourse about women," ultimately argues that the
feminine *esprit* is not inferior to its masculine counterpart in *Marianne*. See
Half-Told Tales, 67–68. In a sense, I agree, in that my argument is not that
Marivaux's feminine *esprit* is less forceful or less efficient than that of men.
However, I would maintain the importance of the fact that its origin (vanity
and petty competition) makes it nonetheless, at best, considerably less re-
spectable than objective, masculine "science." Furthermore, the distinction
between the types of intelligence ascribed to the two sexes is absolute in Marivaux. The highest compliment given to a woman's (Mme. Dorsin) intelligence
in the novel is that of calling it "mâle," thus dispelling any ambiguity about
the status of feminine intelligence. One is reminded of early physiognomical treatises in which "masculine" and "feminine" characteristics transcend
anatomical sex but are clearly considered of different worth.

46. One could, of course, make the point that Marianne's definition is itself
an act of cognition and intellection on the part of a woman.

47. Spitzer, "A propos de *La Vie de Marianne*," 108.

48. In a much more reasonable, thoughtful, and convincing analysis of
Marianne in *The Heroine's Text* (New York: Columbia University Press, 1980),
Nancy K. Miller says that, while she does not agree with "Spitzer's oft-cited
claim," she does read the novel as "feminocentric." She explains: "To the

extent that the feminine—understood as a conventionally assigned and cul-
turally overdetermined attention to the intricacies of sociability and its plots—
is at the heart of his novel, Marivaux is another plausible 'female imperson-
ator' " (151). Miller emphasizes here precisely what Spitzer fails to see—that
the "feminine" is itself a cultural, not biological, phenomenon. Madeleine
Therrien takes a more cynical view of the implications of the character of Marianne: "Marianne est la projection de la chimère masculine: jolie, coquette,
charmante, séduisante . . . L'auteur présente la vision d'une femme à laquelle
on prête des qualités à la condition qu'elle soit jeune et belle. Consciente des
avantages que lui procure la féminité, Marianne se conforme aux demandes
d'une société dominée par des valeurs masculines . . . Le modèle de l'écriture
dans La Vie de Marianne est sans doute féminin, mais l'héroïne est par excellence une création masculine . . ." ("La Problématique de la féminité dans La
Vie de Marianne," Stanford French Review 11, 1 [Spring 1987]: 60). Alfred Cismaru gives a confused but suggestive view of Marivaux's contradictory views
on women, both public and private, in "The Misogyny of Marivaux," Lamar
Journal of the Humanities 11, 1 (Spring 1985): 11–17. Ruth P. Thomas gives a
more coherent argument for misogyny in Marianne, agreeing with Therrien's
view of Marianne as a projection of the male definition of the feminine (which
includes the qualities of intuition and vanity). See Thomas, " 'et je puis dire
que je suis mon ouvrage': Female Survivors in the Eighteenth-Century French
Novel," French Review 60, 1 (October 1986): 7–19.

49. I am of course speculating here about the spectator's possible misogyny, and not Marivaux's; the term "misogynist" would be inaccurate in
reference to Marivaux himself.

50. See Marshall, Surprising Effects of Sympathy, 79–82, for a discussion of
accident as both a philosophical and a grammatical term, as well as the antonym
of essence.

51. Spitzer's contention that "Marivaux ne dédaigne pas la coquetterie" because "la vertu fait feu de tout bois" is not at all convincing, given the origins
of coquetry. See Spitzer, "A propos de La Vie de Marianne," 112.

52. Many think of Marivaux as a sort of protofeminist. It is true that some of
his writings would seem to contradict our assertions, particularly "Réflexions
sur les coquettes" in the fifth feuille of Le Cabinet du philosophe (371–75), in
which he espouses the theory that coquetry is the result of the position of
inferior social power to which women are relegated by men. This definition
of coquetry as a learned defense mechanism indeed contradicts its definition
as an innate, essential quality that we find in Marianne. However compelling,
this stance is contradicted by a number of other positions on the same subject
in the journals (see "Lettres sur les habitants de Paris": "les femmes ont un
sentiment de coquetterie, qui ne désempare jamais leur âme . . . La nature a
mis ce sentiment chez elle à l'abri de la distraction et de l'oubli. Une femme
qui n'est plus coquette, c'est une femme qui a cessé d'être" [28]). See also
L'Indigent philosophe, 280. While these contradictions could certainly constitute
a fascinating study whose aim would be to establish a theory about Marivaux's attitudes toward women in general, they are only of indirect interest

to my work, which is concerned with defining the ideology put forth in *Marianne* alone. For his part, Stewart argues that there are two kinds of coquetry: one natural and innocent, the other artificial and calculating. He characterizes Marianne as the former type of *coquette*, in reference to whom the term is not necessarily a pejorative. See Stewart, *Le Masque et la parole*, 91–93, 140.

53. Peter Brooks makes a similar comment about Emma Bovary: "The reader's sense of the presence of Emma's body is so intense that it comes as something of a surprise, upon rereading the novel, to find that there is really very little in the way of full-length portraiture of Emma." See "The Body in the Field of Vision," *Paragraph* 14, 1 (March 1991): 47.

54. The lack of specific physical description about Marianne is of course to some extent a literary-historical phenomenon: eighteenth- and early-nineteenth-century novels tend to contain little corporeal detail. As I shall discuss in Chapter 4, it was Balzac who introduced detailed physical—and physiognomical—description into the French novel. For an account of this shift in English novels of the mid-nineteenth century and their debt to physiognomy *per se*, see Jeanne Fahnestock, "The Heroine of Irregular Features: Physiognomy and Conventions of Heroine Description," *Victorian Studies* 24, 3 (Spring 1981): 325–50.

55. The two notable exceptions are the brief portraits of Mme. de Fare (237) and the anonymous "parente longue et maigre" of Mme. de Miran (265), both unflattering, almost Dickensian, caricatures of female antagonists.

56. Antoine Furetière, *Dictionnaire universel* (Amsterdam and The Hague, 1690; reprint, Geneva: Slatkine Reprints, 1970).

57. Curiously, Deloffre does not include an entry for *aimable* in his glossaries for either the Garnier Classique edition of *Marianne* or his *Marivaux et le marivaudage*.

58. Given the sexism of the culture in which *Marianne* was written and what Brady tells us about woman as esthetic object in the literature of that culture (see note above), this is hardly a far-fetched notion. Stewart provides yet another meaning of the word as employed by both Marivaux and Duclos, in which *aimable* would describe a person who (more or less intentionally) pleases others. The *coquette*, according to this definition, is necessarily *aimable*. See Stewart, *Le Masque et la parole*, 73, 92–93.

59. Lavater and Marivaux of course share a belief in essentialism common to all physiognomical discourse. Their differences are however more numerous and more enlightening than their similarities.

60. Courtine and Haroche claim (erroneously) that Lavater privileges the pathognomical over the physiognomical. See *Histoire du visage*, 137–44, esp. 138, 144.

Chapter Three. Lavater and *l'alphabet divin*

1. For detailed biographies of Lavater, see Oliver Guinaudeau, *Jean-Gaspard Lavater: études sur sa vie et sa pensée jusqu'en 1786* (Paris: F. Alcan, 1924); Graeme Tytler, *Physiognomy in the European Novel*, 20–34; and Anne-Marie Jaton's re-

cent and handsome *Lavater* in the collection "Les Grands Suisses" (Geneva: Slatkine, 1989).

2. Tytler, *Physiognomy in the European Novel*, 21–22. I use the word "Christocentric" following his example.

3. For information on Lavater's more unorthodox beliefs, see Oliver Guinaudeau, "Au temps du mesmérisme: un fervent adepte du magnétisme animal, le pasteur J.-G. Lavater," *Etudes germaniques* 13, 2 (1958): 98–113.

4. Tytler, *Physiognomy in the European Novel*, 30–32. For a contemporary polemic against Lavater's Jesuit sympathies and mysticism, see Honoré Gabriel Riqueti, comte de Mirabeau, *Lettre à M . . . . sur M.M. Cagliostro et Lavater* (Berlin: François de la Garde, 1786).

5. For the following information about the publishing of the *Fragmente* and its translations, I am largely indebted to Guinaudeau, *Jean-Gaspard Lavater*, 208ff. and Tytler, *Physiognomy in the European Novel*, 56–60 and 82–84.

6. For pertinent bibliographical references on the collaborative aspect of the *Fragmente*, see Tytler, *Physiognomy in the European Novel*, 339, note 85.

7. Guinaudeau, *Jean-Gaspard Lavater*, 251.

8. Tytler, *Physiognomy in the European Novel*, 57 and Guinaudeau, *Jean-Gaspard Lavater*, 251.

9. Guinaudeau, *Jean-Gaspard Lavater*, 209.

10. The translators listed in the original French edition are "Mme. E. de la Fîte, H. Renfner, et Mlle. Caillard." Tytler tells us (*Physiognomy in the European Novel*, 346, note 6) that Mme. de la Fîte was probably the most important contributor to the edition, and the wife of the minister of the Walloon community in The Hague.

11. Quoted in Tytler, *Physiognomy in the European Novel*, 346, note 6. Guinaudeau tells us: "Lavater ne fera plus de l'édition française une simple traduction de l'allemande, mais un nouveau travail original . . . ," *Jean-Gaspard Lavater*, 611, note 285. See also the translator's preface to the original French edition, in which he asserts that the edition is not merely a translation from the German, but also includes original material.

12. Although it is not the primary object of this study to try to define the direct influence of Lavater's work on French literature, I will nonetheless be suggesting relations between certain works of French literature and physiognomical thought (including Lavater's). These relations do include, at least anecdotally, such facts as our knowledge that Balzac did indeed read Lavater (in French translation). It therefore seems appropriate in this context that I use the original, expanded French version of Lavater's work as my primary text.

13. Tytler, *Physiognomy in the European Novel*, 84.

14. It is specifically his conflation of physiognomy and theology that rendered Lavater's work unique. The more general conflation of theology and science was an eighteenth-century commonplace. For titles of works which best illustrate this phenomenon, see Tytler, *Physiognomy in the European Novel*, 53–54.

15. It is easy to recognize the buds of physiognomical thought in some of Lavater's earlier, more explicitly theological works, particularly *Ausssichten in*

die Ewigkeit (1768–73), in which he discusses such related topics as the relation between body and soul, and the bodies and language of angels. See
Guinaudeau, *Jean-Gaspard Lavater*, 122–23.

16. For further information on the close relation between Lavater's "Christocentric" theology and his physiognomy, see Guinaudeau, *Jean-Gaspard
Lavater*, 147–48, and Stafford, *Body Criticism*, 91–93. See Guinaudeau, 581,
note 161 for a reference to Lavater's rather bizarre notion that Jesus Christ, as
the "ideal man," must also have been the most beautiful.

17. The subtitle, *zur Beförderung der Menschenerkenntniss und Menschenliebe*,
was more or less literally translated into French. The reason for the choice of
"essai" does not seem to have been documented.

18. I am using here the somewhat anachronistic twentieth-century definition of the term "science" by which to make my judgments. In spite of
the vast difference between eighteenth- and twentieth-century criteria for the
establishment of a science (the elaboration of which would constitute another
study entirely), criticism of Lavater by some of his contemporaries attests to
his failure to conform to their standards as well as to ours.

19. The equation here of physiognomy with astrology and chiromancy
(palmistry) is not an arbitrary one, as they have often been associated with
each other. In fact, at one point in British history, anyone found "guilty" of
practicing any one of the three was subject to decapitation. All three have
their roots in divination going back to Babylonia. See Bottéro, *Divination et
rationalité*, 70–196.

20. Barbara Stafford neatly sums up the impossibility of any practical application of Lavater's rather random observations as follows: "Speculative bits of
wisdom were dished out independent of any practical existential situation."
She goes on to characterize his work as "a dissonant pictorial and linguistic
parade." See *Body Criticism*, 102.

21. Although my own reading of Lavater will thus take as its object the
(verbal) rhetoric of the text, it is at least equally valid to approach the *Essai* primarily through its pictorial elements, which constitute their own (nonverbal)
language and rhetoric and which, in their nature as "mosaics," are related to
the fragmentary verbal text itself. Indeed, this is the task Barbara Stafford sets
herself in her analysis of Lavater—the result is perhaps the most provocative,
and certainly the most original, treatment of Lavater to date. See *Body Criticism*, esp. 91–103. There are numerous references to Lavater throughout *Body
Criticism*.

22. In his positivist theory of language, Auguste Comte posits a similar idea
of an original, biological language which precedes artificial, human language.
See *Système de politique positive* (Paris: Chez Carilian-Goeury et V. Dalmont,
1851–54), 2:216–62. The eighteenth-century English philosopher Berkeley also
articulates a distinction between a divine, nonverbal language and human
language. See *The Works of George Berkeley* (London: Thomas Nelson and Sons,
1948), 1:251–76. In the course of her examination of the pictorial elements
of Lavater's *Essai*, Barbara Stafford makes an astute and unexpected link between the visual and the verbal by remarking that the very layout of Lavater's

pages and form of the portraits they contain invite "an etymological approach to the face." She goes on to describe "six unrelated portraits of clerics" as "paratactically juxtaposed like verbless nouns." See *Body Criticism*, 95.

23. For Saussure's definition of "signifié" and "signifiant," see Ferdinand de Saussure, *Cours de linguistique générale* (Paris: Payot, 1976), 98–100.

24. See G. P. Brooks and R. W. Johnson, "Johann Caspar Lavater's *Essays on Physiognomy*," *Psychological Reports* 46 (1980): 10–12, for their observation that Lavater's use of terms to describe psychological states is "neither defined nor classified" and that "he was actually content . . . to do little more than name such traits of personality as benevolence, brutality, or voluptuousness" (12). See John Locke, *An Essay concerning Human Understanding* (New York: Dover Publications, 1959), 2:152 and 157–58, on the necessity of defining abstract terms.

25. Condillac, *Essai sur l'entendement humain* (Paris: Librairie Armand Colin, 1924), 32. All subsequent quotes from this edition.

26. Rousseau, *Essai sur l'origine des langues* (Paris: Bibliothèque du Graphe, 1970), 504. All subsequent quotes from this edition.

27. Rousseau's endorsement of imagistic, descriptive language ties in with passages from both Lavater and Condillac which we will see below.

28. See Berkeley for the idea that a God-given, natural language and an artificial, human language can be equally arbitrary.

29. Eighteenth-century French literature is full of what might be called "pathognomical novels" in which much of the intrigue centers on the concealing, revealing, and deceiving possibilities of pathognomy. Two novels exemplary of these obsessions are Crébillon's *Les Egarements du cœur et de l'esprit* and Laclos's *Les Liaisons dangereuses.*

30. For a discussion of the various manifestations within the *Essai* of Lavater's "anatomical preoccupation with the paralyzed, stiff, and cold signs of death," see Stafford, *Body Criticism*, 101–2. The emphasis on the dimensions of the forehead and skull were of course later developed into the even more scientistic discipline of phrenology by Franz Josef Gall. Phrenology and physiognomy are frequently confused. For a discussion of the relation between the two, and specifically Gall's debt to Lavater, see Brooks and Johnson, "Lavater's *Essays on Physiognomy*."

31. *Oeuvres philosophiques de Buffon* (Paris: PUF, 1954), 298. Subsequent references are to this edition.

32. See ibid., 299–304. The passage on the communicative powers of the human eye is on 299.

33. Both the *Encyclopédie* and the *Grand Dictionnaire du XIXe siècle* quote a version of the passage below in their entries for "physiognomy."

34. Many more recent scholars have studied facial expression, including Darwin and, at present, Ray Birdwhistell. Surprisingly, the study of the fixed features as an index to character (now most often referred to as "morphopsychology") lives on as well; as recently as 1980, a volume of the *Que sais-je?* series entitled *Physionomie et caractère* was in its tenth edition. Francis Baud, *Physionomie et caractère* (Paris: PUF, 1980). For even more mainstream contem-

porary works on physiognomy, see Timothy T. Mar, *Face Reading* (New York: Signet Books, 1974); Jean Lefas, *Physiognomy: The Art of Reading Faces* (Ariane Books, 1975); and Terry Landau, *About Faces* (New York: Doubleday, 1989). It is interesting to note that Landau's very recent publication bears a subtitle which reveals age-old physiognomical concerns: "The Evolution of the Human Face: What It Reveals, How It Deceives, What It Represents, and Why It Mirrors the Mind."

35. Guinaudeau gives the following biographical explanation of Lavater's denial of the arbitrary: "Affirmer, en effet, que les traits de chaque créature, les moraux comme les physiques, ont un caractère individuel, "arbitraire," séduisait certes Lavater, tempérament individualiste, ami de plusieurs "Stürmer"; mais cette théorie paraissait condamner d'avance sa prétention d'établir une science physiognomonique: il n'est, en effet, de science que du général. Soucieux de sauver la nouvelle science, Lavater bannit donc tout arbitraire . . . des *Fragments;* il ne veut pas de cette notion dans la 'saine philosophie de la nature' et, par conséquent, ne l'admet pas dans la Physiognomonie" (*Jean-Gaspard Lavater*, 271–72).

36. Saussure, *Cours de linguistique générale*, 101.

37. A perhaps equally compelling study of Lavater could be undertaken from the perspective of art criticism (both analyzing the images incorporated in Lavater's work and criticizing his readings of them); this, however, would constitute an entirely different study and is not necessarily germane to a rhetorical study of physiognomy and literature such as mine.

38. See Brooks and Johnson, "Lavater's *Essays on Physiognomy*," 10–12, and Locke, *Essay concerning Human Understanding*, 157–58.

39. This passage comes from fragment 18, entitled "Observations d'un Savant Allemand sur la Physiognomonie." The "savant" in question is never identified, and the form of the fragment, that of a dialogue between Lavater and the "savant," makes it extremely difficult to ascertain who is saying what.

40. Lavater poses questions in this passage which were also important concerns for the Romantics and, later, for Nietzsche.

41. For a discussion of the generality of language, see A. B. Johnson, *A Treatise on Language* (New York: Dover Publications, 1968), Part First, lecture 7, 114–15. See also Locke, *Essay Concerning Human Understanding*, 2:14–31 (chap. 3) and 148–64 (chap. 11).

42. The distinction between language in its concretely referential usage and as used for the purposes of abstraction is made, incidentally, by Condillac: "L'usage est uniforme et constant pour les noms des idées simples, et pour ceux de plusieurs notions familières au commun des hommes; alors il ne faut rien changer; mais, lorsqu'il est question des idées complexes qui appartiennent plus particulièrement à la métaphysique et à la morale, il n'y a rien de plus arbitraire, ou souvent de plus capricieux . . ." (199). "Pour la signification des noms des idées simples, qui viennent immédiatement des sens, elle est connue tout à la fois . . . Ces sortes de termes ne peuvent donc être obscurs. Le sens en est si bien marqué par toutes les circonstances où

nous nous trouvons naturellement, que les enfans mêmes ne sauroient s'y tromper" (201).

43. See Locke, *Essay concerning Human Understanding*, 2:152 and 157–58.

44. For theoretical discussions of the proper name, see *La Logique de Port Royal*, part 1, chap. 6, quoted in Alain Rey, ed. *Théories du signe et du sens* (Paris: Klincksieck, 1973), 1:115–16, and John R. Searle, "Proper Names," *Mind* 47 (1958):166–73.

45. A direct refutation of this belief in a correspondence between the material and the immaterial can be found in Buffon's critique of physiognomy, cited above.

46. Michael Shortland, "Skin Deep: Barthes, Lavater and the Legible Body," *Economy and Society* 14 (1985):285.

47. Ibid.

48. Ibid., 286.

49. Ibid.

50. Ibid., 291.

51. Ibid., 294.

52. This notion has obvious ties to the German pre-Romantic "Sturm und Drang" idea of genius, as Guinaudeau states: "Les Stürmer affirment que, seule, l'intuition nous révèle la vérité, qui est, par conséquent, toute subjective; selon eux, l'intuition, le pressentiment tiennent lieu de toutes les règles de la logique. Lavater est engoué de théories analogues . . . le 'génie' ne connaît point la vérité par le raisonnement, par l'enchaînement logique des causes et des effets" (*Jean-Gaspard Lavater*, 335).

53. As Guinaudeau tells us in a passage outlining some of the contradictions in Lavater: "N'était-il pas . . . étrange et contradictoire de voir notre pasteur . . . tout à la fois poursuivre des recherches qu'il répute scientifiques et, dans l'un de ses 'Discours,' souhaiter au physiognomiste une pointe d'imagination qui lui permette d'opérer entre les visages les rapprochements nécessaires?" (ibid., 211).

54. As I have mentioned above, Barbara Stafford expresses, more elegantly than I, a similar idea concerning the practical inadequacy of Lavater's *essai*. See my note 20.

55. Mirabeau, *Lettre à M . . . . sur M.M. Cagliostro et Lavater*, 27.

56. For a catalogue of these successors in both France and England as well as a suggestive discussion of their influence on the Victorian novel, see Fahnestock, "Heroine of Irregular Features," esp. 336–39. See also Stafford, *Body Criticism*, 103–20.

57. Chateaubriand, in a note to his *Essai sur les révolutions* (part 1, chap. 19), makes the same rapprochement between this café sitter's game and Lavater's physiognomy: "Toute l'antiquité a cru à la vérité de cette science, et Lavater l'a portée de nos jours à une perfection inconnue. La vérité est que la plupart des hommes la rejettent parce qu'ils s'en trouveraient mal . . . J'aime à aller m'asseoir, pour ces espèces d'observations, dans quelque coin obscur d'une promenade publique, d'où je considère furtivement les personnes qui pas-

sent autour de moi . . ." The passage is quoted by Fernand Baldensperger, "Les Théories de Lavater dans la littérature française," in his *Etudes d'histoire littéraire* (Paris, 1910; Geneva: Slatkine Reprints, 1970), 66.

58. Brooks, *Reading for the Plot*, 10; Brooks develops this idea of narrative and its relation to "the organization of biography" (320) further on in his study, particularly in reference to Freud and his "masterplot" (chap. 4).

59. Brooks quotes Paul Ricoeur's definition of plot as "the intelligible whole that governs a succession of events in any story" (ibid., 13), a definition well in keeping with his own statements on plot and narrative.

60. Ibid., 6.

61. Ibid., 5–6.

62. In an interesting moment in the preface to the *Essai*, Lavater states a goal for his own work more usually associated with fictional narrative, the creation and sustaining of interest ("desire," if you will) by the text in its reader. "j'ose promettre . . . d'employer tous mes efforts pour que l'intérêt du livre aille toujours en croissant . . ." (1:viii, préface).

63. For a theoretical discussion of description and its relation to narration, see Philippe Hamon, *Introduction à l'analyse du descriptif* (Paris: Hachette, 1981), esp. 42–51. Hamon's analysis of description in the novel will be crucial to my readings of specific novels in later chapters.

64. Peter Brooks refers to this basic "narrative impulse" in the first sentences of *Reading for the Plot:* "Our lives are ceaselessly intertwined with narrative, with the stories that we tell and hear told, those we dream or imagine or would like to tell, all of which are reworked in that story of our own lives that we narrate to ourselves in an episodic, sometimes semiconscious, but virtually uninterrupted monologue. We live immersed in narrative, recounting and reassessing the meaning of our past actions, anticipating the outcome of our future projects, situating ourselves at the intersection of several stories not yet completed. The narrative impulse is as old as our oldest literature . . ." (3).

65. Ibid., 5. It is pertinent to make the point here that Lavater could be seen as an amalgam of the Enlightenment and Romanticism, in his contradictory pretentions to, on the one hand, "cold," objective observation and, on the other, emotive, subjective "genius." At one point, Brooks makes a very suggestive analogy between "plot" as a dynamic of narrative and the term "plotting points" as it is used in geometry to refer to the inscription of a curve or other figure in a plane.

66. Physiognomical description in the novel creates in part what Philippe Hamon calls the "effet-personnage." See Hamon, *Introduction à l'analyse du descriptif*, 110–17.

67. See Courtine and Haroche, *Histoire du visage*, 140.

Chapter Four. Balzac and *l'homme hiéroglyphié*

1. See Fernand Baldensperger, *Orientations étrangères chez Balzac* (Paris: Honoré Champion, 1927), 82, among other critics. Judith Wechsler, in *A Human Comedy: Physiognomy and Caricature in Nineteenth-Century Paris* (Chi-

cago: University of Chicago Press, 1982), says that "there is some disagreement as to which edition [of Lavater] Balzac owned . . ." (182).

2. Pierre Abraham, *Créatures chez Balzac* (Paris: Gallimard, 1931), 117.

3. Quoted by Baldensperger, *Orientations*, 84.

4. See Baldensperger, "Les Théories de Lavater dans la littérature française," 84. Also cited by Tytler, *Physiognomy in the European Novel*, 4. Balzac has some competition for this title of "most physiognomical" of the nineteenth-century European novelists, particularly from the English. For an excellent account of the widespread influence of Lavaterian physiognomy and its later spin-offs on the Victorian novel, see Fahnestock, "Heroine of Irregular Features."

5. Abraham, *Créatures chez Balzac*, 117.

6. See also Baldensperger, "Les Théories de Lavater," 83–84, for the influence of Lavater on Balzac's theory of the body as an ensemble, as well as Gilbert Malcolm Fess, *The Correspondence of Physical and Material Factors with Character in Balzac* (Philadelphia: Publications of the University of Pennsylvania, 1924), 57–58.

7. The above passages from Lavater and Balzac are quoted by Baldensperger, 87–88.

8. See Fess, *Correspondence*, 40–60, on correspondences between Lavater and Balzac.

9. See ibid., 63–70, for a catalogue of these Balzacian contributions to physiognomical thought.

10. See ibid., 77–98. See also Baldensperger, *Orientations*, 89–91, on Balzac's "exaggeration" of Lavater's suggestions about *milieu*. In his obsession with *milieu*, Balzac is in a very significant sense Zola's precursor.

11. Fess, *Correspondence*, 61–62.

12. See Abraham, *Créatures chez Balzac*, 119–20.

13. Quoted by Baldensperger, *Orientations*, 93.

14. Henri Gauthier, *L'Image de l'homme intérieur chez Balzac* (Geneva: Droz, 1984), 119.

15. Gauthier exposes an elaborate set of relations between Mesmerism and Swedenborgianism in Balzac, thematized through a number of angel-like characters he calls "angélomorphes." See Gauthier's third chapter, "L'Anthropologie angélique," *L'Image de l'homme intérieur*, 159–251. Because it is an enlightening theoretical analogue to physiognomy, and not merely a variant of it, the "vital fluid" theory is obviously of considerably more interest to my discussion than would be phrenology.

16. Balzac, Avant-propos to *La Comédie humaine*, in *Oeuvres complètes* (Paris: Furne, Durochet, et al., 1842), 8–9. All subsequent references will be to this edition and be identified simply as "Avant-propos".

17. Baldensperger, *Orientations*, 75.

18. Peter Demetz, "Balzac and the Zoologists: A Concept of the Type," in *The Disciplines of Criticism: Essays in Literary Theory, Interpretation, and History* (New Haven: Yale University Press, 1968), 409. Demetz goes on to say that we shouldn't "press the point too far" (about Balzac's man as a product

of a historical milieu) as it was actually Hippolyte Taine, in the 1860s, who transformed Saint-Hilaire's *espace, temps, et lieux* into his own *race, moment, et milieu*.

19. Wechsler provides a history of the *Physiologies* and their immense popularity: "Balzac's vignettes on social types developed into an independent semi-journalistic genre—the *Physiologies*—which, under the leadership of the editor Philipon, was taken up by other writers and illustrated by leading draughtsmen of the day. This literature began in the 1830's, reached its peak in the early 1840's and subsided in the 1850's . . . The descriptive *Physiologies* . . . flooded the bookstalls between 1841 and 1843. Their vogue in these years was spectacular . . . Balzac claimed paternity of the *Physiologies* as a genre . . . Half a million copies of the *Physiologies* were bought during the few years of their vogue . . ." (*Human Comedy*, 31–34).

20. It is interesting to note that the appeal of such popular sociological treatises has experienced something of a renaissance in the past decade or so in the United States, with such comic works as Paul Fussell's *Class* (New York: Ballantine Books, 1983) and Mike Dowdall and Pat Welch's *Humans* (New York: Simon and Schuster, 1984). Even more widely known are Lisa Birnbach's *The Official Preppy Handbook* (New York: Workman Publishing, 1980) and its counterpart, Marissa Piesman and Marilee Hartley's *The Official Yuppie Handbook* (New York: Pocket Books, 1984). Interestingly, the French have reappropriated the genre with their own version of the *Preppy Handbook,* Thierry Mantoux's *BCBG: Le Guide du bon chic bon genre* (Paris: Seuil, 1985). These books produce much the same results that Balzac's *Physiologies* probably did: a comic effect resulting from the tension between the overtly parodical form and the sometimes undeniably accurate descriptive content.

21. See the entry "sociology" in *The New Columbia Encyclopedia* (New York: Columbia University Press, 1975), 2553, for a brief but authoritative history of the discipline. Ruth Amossy gives a very insightful reading of the concepts of *type* and *stereotype* in relation to the genre of the *physiologie*. Perhaps the most important element of her discussion is the idea that the critique of the *physiologie*, widespread at the time of their popularity, was based not on the notion of typology itself, but rather on the sloppy form and style of the mass-produced pamphlets. It was not until the twentieth century that the notions of the usefulness and even possibility of typology were put into question. Nineteenth-century urban culture held to this kind of classificatory thinking as a necessary (and reassuring) means of making sense of new sociological patterns and denying the seemingly arbitrary nature of an anonymous society. The *physiologie* is thus a particularly justified taxonomy. See Amossy, "Types ou stéréotypes? Les *Physiologies* et la littérature industrielle," *Romantisme* 18, 64 (1989): 113–23.

22. From *Traité de la vie élégante* in Balzac, *Les Parisiens comme ils sont*, ed. Andre Billy (Geneva: La Palatine, 1947), 189. All subsequent references to the *Traité* will be to this edition.

23. There are a number of articles on Balzac and vestiognomy. They include the following: Jeanne Reboul, "Balzac et la vestiognomie," *Revue d'histoire*

littéraire de la France 50, 2 (Apr.–June 1950): 210–33; Mireille Labouret-Grare, "L'Aristocrate balzacienne et sa toilette," *L'Année balzacienne* 3 (1982): 181–93; and Danielle Dupuis, "La Poésie de la toilette féminine chez Balzac," *L'Année balzacienne* 5 (1985): 173–95. As we shall see below, vestiognomy plays an important role in *La Vieille Fille*. Fredric Jameson discusses clothes as sociopolitical indices in "The Ideology of Form: Partial Systems in *La Vieille Fille*," *Sub-Stance* 15 (1976): 29–49.

24. The usage of the term "physiology" we accept today was not yet standardized in Balzac's time. It can probably most accurately be said to have become standard with Claude Bernard's definition of it in the latter half of the nineteenth century.

25. Peter Brooks, "The Text of the City," *Oppositions* 8 (1977): 10. Brooks goes on to suggest the extremely far-reaching epistemological implications of this Balzacian mode of thought: "He [Balzac] invents the nineteenth-century by bringing to consciousness the very shape of modernity as a set of texts subject to our reading and interpretation" (10).

26. Balzac, *Traité de la vie élégante, suivi de la Théorie de la démarche*, ed. Claude Vareze (Paris: Bossard, 1922), 123–24. All subsequent citations are from this edition.

27. Wechsler, *Human Comedy*, 30–31.

28. Martin Kanes, in *Balzac's Comedy of Words* (Princeton: Princeton University Press, 1979), suggests the possibility of a historical relation between the two texts: "The *Traité de la vie élégante* mentions a future chapter on the theory of deportment, which possibly becomes the essay entitled *Théorie de la démarche*" (159).

29. In his article "Balzac, Précurseur de la caractérologie," *L'Année balzacienne* (1965), Paul Metadier remarks that, in the *Théorie*, Balzac "voulait et trop de science et trop de frivolité peut-être. Placer la science dans un décor de frivolité, rien ne convenait mieux à Balzac installé dans son poste d'observation du boulevard des Italiens" (310). He goes on to give the following assessment of the seriousness of the text: "*La Théorie de la démarche* n'est pas une simple fantaisie, une manifestation négligeable de l'esprit boulevardier. Toute la faculté d'observation de Balzac s'y donne libre cours. Elle sait découvrir les facteurs psychologiques qui animent les pantins de la société parisienne que nous voyons défiler sous ses yeux" (311).

30. For an abstract rhetorical reading of tautology as the central rhetorical figure in the *Théorie*, see William R. Paulson, "La Démarche balzacienne ou le livre sur rien," *Romanic Review* 75, 3 (May 1984): 294–301.

31. *Le Charivari*, October 25, 1836. Quoted by P. G. Castex in his introduction to *La Vieille Fille* (Paris: Garnier, 1957), xxxiv–xxxv. All subsequent citations from *La Vieille Fille* will refer to this edition as well.

32. Georges Laffy, in "La Politique dans *La Vieille Fille*," *Ecrits de Paris*, no. 297 (November 1970), gives a concise version of this reading of the novel: "On peut donc voir la fantaisie du livre *La Vieille Fille* comme la démonstration d'un théorème qu'on pourrait énoncer ainsi: la monarchie et la république se disputent, sous la Restauration, la province française. Si, à la fin, la

république l'emporte, c'est pour le plus grand malheur de la province" (67). Fredric Jameson gives the most complex and suggestive reading of the novel as political allegory in "The Ideology of Form."

33. Reboul, "Balzac et la vestiognomie," 224.

34. Bernard Vannier, *L'Inscription du corps: pour une sémiotique du portrait balzacien*, (Paris: Klincksieck, 1972), 20. Vannier identifies this phenomenon of "personnages contrastes" as one which appears in various novels by Balzac.

35. In *L'Expression métaphorique dans la Comédie humaine: domaine social et physiologique* (Paris: Klincksieck, 1976), Lucienne Frappier-Mazur defines monomania in the Balzacian universe: "La monomanie est le point extrême de la concentration énergétique et cette image dépeint exclusivement diverses formes d'idée fixe—amour, avarice . . . La monomanie . . . se présente comme un transfert d'énergie sexuelle" (233, 238). Frappier-Mazur also points out that women, children, and "le sauvage, bête sauvage ou criminel" are "trois êtres chez qui domine l'instinctivité" and are therefore most likely to be mono-maniacal; she notes that in the *Comédie* women are portrayed as having "élé-ments disparates" but that their "volupté" provides a "courant unificateur" (236–37).

36. Whether there is in fact any such physiological relation is beside my point; I am merely exposing the beliefs on which the novel functions.

37. Less central to the plot of the novel, but equally demonstrative of Mlle. Cormon's blindness to physiognomical signs, is her complete failure to rec-ognize that Athanase Granson is in love with her ("Il tenait encore . . . une passion qui lui creusait les joues et lui jaunissait le front"). This failure leads, indirectly, to his tragic suicide. See Jeannine Guichardet, "Athanase Gran-son, corps tragique," *L'Année balzacienne* 5 (1985): 151–60, on Mademoiselle Cormon's misreading of Athanase and how this leads to his death.

38. For a study on these particular constructions in Balzac, see Ian Pickup, "An Aspect of Portraiture in the Novels of Balzac: Psycho-Physiological Bridges," in *Literature and Society: Studies in Nineteenth and Twentieth Century French Literature*, ed. C. A. Burns. (Birmingham, England: Goodman for Uni-versity of Birmingham, 1980), 38–55.

39. It is important to note here the inescapable contradiction between this definition of physiognomical perception and the narrator's earlier lament that Mlle. Cormon's illiteracy could have been solved by the presence in the re-gion of a professor of "anthropologie." Balzac seems to suggest that physiog-nomical skill is both innate and acquired. In so doing, he—perhaps uncon-sciously—mirrors one of the most important rhetorical failures of Lavater's *Essai*. For a detailed discussion of the paradox in Lavater, see Chap. 3.

40. Jameson, "Ideology of Form," 38. Jameson comments here on the end-ing of the novel in the context of a larger discussion of its narrative dynamic.

41. The introduction to *La Fille aux yeux d'or* is indeed a sort of *physiolo-gie*. See Jean-Yves Debreuille, "Horizontalité et verticalité: inscription idéo-logiques dans *La Fille aux yeux d'or*," in *La Femme au XIXe siècle: littérature et idéologie* (Lyon: Presses Universitaires de Lyon, 1979), 151.

42. Michel Nathan, "Zoologies parisiennes," in *La Femme au XIXe siècle*, 191–92.

43. At the end of the century, the title character of Oscar Wilde's *Salome* (1891) will also have golden eyes. This is particularly worthy of note in light of the fact that she is of course, like Paquita, a femme fatale.

44. For a detailed analysis of the use of this appellation in the story, see Maurice Laugaa, "L'Effet *Fille aux yeux d'or*," *Littérature* 20 (December 1975): 62–80.

45. On the conventions of Balzacian beauty and their literary origins, see Roland Le Huenen, "L'Ecriture du portrait féminin dans *La Cousine Bette*," in *Balzac et les parents pauvres*, ed. Françoise van Rossum-Guyon and Michiel van Brederode (Paris: SEDES, 1981), 78. See especially Vannier, *L'Inscription du corps*, 138.

46. For a discussion of the Freudian notions of an eroticized desire to know ("epistomophilia") and its relation to the equally eroticized desire to look ("scopophilia"), see Peter Brooks, "The Body in the Field of Vision," *Paragraph* 14, 1 (March 1991), and Janet Beizer, "The Body in Question: Anatomy, Textuality, and Fetishism in Zola," *L'Esprit créateur* 29, 1 (Spring 1989): 50–60.

47. Pierre Abraham, *La Figure humaine chez Balzac: la couleur des cheveux* (Paris: Bulletin de la Société de la Morphologie, 1928).

48. In the most convincing and suggestive article to date on *La Fille aux yeux d'or*, "Rereading Femininity," *Yale French Studies* 62 (1981), Shoshana Felman gives a brilliant reading of the psychosexual implications of the gold eyes in which the gold of the eyes, as a "reflective substance," serves to reflect Henri's own desire. It therefore, according to Felman, defines the feminine as a reflection of the masculine: "Defined by man, the conventional polarity of masculine and feminine names woman as a *metaphor of man*. Sexuality, in other words, functions here as the sign of a rhetorical convention, of which woman is the *signifier* and man the *signified*. Man alone thus has the privilege of proper meaning, of *literal* identity; femininity, as signifier, cannot signify *itself*; it is but a metaphor . . ." (25).

49. As Ruth Amossy says with respect to the classificatory discourse of the *physiologie*: "le type ramasse lui aussi en une image unifiée et nécessairement simplifiée les attributs caractéristiques d'un groupe social. Voué à la généralisation, il ne s'attache pas aux nuances du cas individuel." See "Types ou stéréotypes?" 114.

50. Balzac, *La Fille aux yeux d'or* in *La Duchesse de Langeais, suivi de La Fille aux yeux d'or*, ed. Pierre Barbéris (Paris: Le Livre de Poche, 1972), 258. All subsequent citations refer to this edition.

51. Serge Gaubert, "*La Fille aux yeux d'or*: un texte-charade," in *La Femme au XIXe siècle*, 169.

52. Felman, "Rereading Femininity," 38–39.

53. *Signes trompeurs*, physiognomical and other, are a common device in the early *roman policier*. The first example which comes to mind is Balzac's *Une Ténébreuse Affaire*. See David Miller, *The Novel and the Police* (Berkeley: University of California Press, 1987).

54. In *The Melodramatic Imagination,* Peter Brooks says that "in Balzac's most ambitious moments" "the direct articulation of central meanings is difficult, dangerous, and even impossible." "But," he continues, "this is not viewed as reason to abandon the search for them" (199). *La Vieille Fille* and *La Fille aux yeux d'or* are clearly, in this sense at least, among Balzac's "most ambitious moments."

55. The term *effet de réel* was coined by Roland Barthes in his essay on the function of seemingly arbitrary descriptive detail in "realist" narrative, "L'Effet de réel," in *Littérature et réalité* (Paris: Editions du Seuil, 1982), 81–90. Barthes's basic theory is that such details serve to further the illusion of reality which a certain genre of fictional narrative tries to create.

56. Vannier, *L'Inscription du corps,* 184. Charles Grivel, in "L'Histoire dans le visage," in *Les Sujets de l'écriture,* ed. Jean Decottignies (Lille: Presses Universitaires de Lille,1981), echoes this idea: "*Le travail narratif du visage.* Le visage permet d'insérer dans le récit de quoi fonder son cours, point n'est besoin d'insister" (195).

57. See Vannier, *L'Inscription du corps,* 18.

58. Brooks, *Melodramatic Imagination,* 125. For a related discussion, see Gerard Genette, "Vraisemblance et motivation," in *Figures II* (Paris: Editions du Seuil, 1969).

59. The similarity of the Balzacian and Lavaterian agenda of denying the arbitrary is striking.

60. The term is from Brooks, *Melodramatic Imagination,* 126.

61. See Wolfgang Iser, *The Act of Reading: A Theory of Aesthetic Response* (Baltimore: Johns Hopkins University Press, 1978), ix–xii. For the present purpose, I have obviously simplified a few of the rather abstract points of Iser's theory of the reception of literary texts.

62. Ibid., 59.

63. Ibid., 48.

Chapter Five. Gautier and *la beauté de l'ambigu*

1. For a presentation of the contemporary reception of the novel, see René Jasinski, *Les Années romantiques de Théophile Gautier* (Paris: Vuibert, 1929), 322–26. See also Janet Sadoff, "Ambivalence, Ambiguity, and Androgyny in Théophile Gautier's Mademoiselle de Maupin," Ph.D. diss., Harvard University, 1987, 1–20.

2. Among many others, see Jasinski, *Les Années romantiques,* and Rita Benesch, *Le Regard de Théophile Gautier* (Zürich: Juris Druck Verlag, 1969). At one point, Benesch typifies the biographical fallacy in criticism on *Mademoiselle de Maupin* by introducing a sentence with the phrase "Dans *Mademoiselle de Maupin,* Gautier avoue que, tout enfant, son plus grand plasir était . . ." (35). The confusion of Gautier the novelist with d'Albert the fictional character is a commonplace, as is Benesch's conclusion: "Gautier s'est donc incarné dans ses personnages par un certain côté de sa personnalité" (94). Indeed, Benesch identifies *Mademoiselle de Maupin* as Gautier's "autobiographie intérieure de

sa jeunesse" (17). For a refutation of the biographical fallacy in Gautier criticism, see Richard B. Grant, *Théophile Gautier* (New York: Twayne Publishers, 1975), 35–36. Grant in turn cites Albert Smith as "the only critic to sense that this approach can be abused" (35–36). See Smith, *Ideal and Reality in the Fictional Narratives of Théophile Gautier* (Gainesville: University of Florida Press, 1969). Rosemary Lloyd gives us the most articulate caveat about biographical readings of *Mademoiselle de Maupin* and the temptation to conflate author and "protagonist": "It is part of the novel's challenge to the reader that it abounds in hints that Gautier, through gentle mockery and irony, is distancing himself from the male protagonist . . . the gap between what d'Albert perceives to be the truth . . . and what the reader sees to be true is too great for there to be any possibility that d'Albert speaks constantly and consistently for Gautier." See Lloyd, "Rereading *Mademoiselle de Maupin*," *Orbis Litterarum* 41, 1 (1986): 19–32.

3. Anne Bouchard, "Le Masque et le miroir dans *Mademoiselle de Maupin*," *Revue d'histoire littéraire de France* 72, 4 (July–Aug. 1972): 584.

4. Ibid., 599.

5. The most notable criticism on *Mademoiselle de Maupin* (or at least that which I have found most suggestive) is relatively recent and includes Rosemary Lloyd, "Rereading *Mademoiselle de Maupin*," and "Speculum Amantis, Speculum Artis: The Seduction of *Mademoiselle de Maupin*," *Nineteenth-Century French Studies* 15, 1–2 (Fall–Winter 1986–87): 77–86; and Kari Weil, "Romantic Androgyny and Its Discontents: The Case of *Mademoiselle de Maupin*," *Romantic Review* 78, 3 (May 1987): 348–58. Lloyd gives a succinct and accurate refutation of the various fallacies that have plagued the criticism on *Mademoiselle de Maupin* in "Rereading *Mademoiselle de Maupin*," 20–21. Interestingly, an inordinate number of American Ph.D. dissertations on *Mademoiselle de Maupin* were produced in the period 1985–89: see Joan Marie Boczenowski, "Art and Nature in *Mademoiselle de Maupin* and *A Rebours*," University of Virigina, 1985; Nathalie David-Weill, "La Femme dans l'oeuvre de Théophile Gautier: de la description à l'interprétation," New York University, 1988; Helga Druxes, "The Feminization of Doctor Faustus: The Quest for Self-Knowledge in the Nineteenth Century," Brown University, 1987; Jacqueline Klaasen, "A Figural Model for Describing the Interpretation of a Novel: Theophile Gautier's *Mademoiselle de Maupin*," University of Wisconsin-Madison, 1986; Sadoff; and Kari Weil, "Veiling Desire: The Aesthetics of Androgyny in Balzac and Gautier," Princeton University, 1985. One might indeed wonder if the novel is slowly making its way from the margins into the mainstream of the nineteenth-century canon, if there is indeed a reversal of what Serge Fauchereau calls "le traitement négligé réservé à Gautier . . . l'aboutissement d'une longue tradition de mépris à son égard." See Fauchereau, *Théophile Gautier* (Paris: Editions Denoël, 1972), 14. The work of Boczenowski, Weil, and especially Sadoff has been essential to my own study of the novel.

6. A recent and undeniably compelling discussion of such questions (the distinction between sex and gender, the construction of the binary opposition between the two sexes, the organizing principle of "compulsory heterosexu-

ality") is to be found in Judith Butler, *Gender Trouble: Feminism and the Subversion of Identity* (New York: Routledge, 1990). The most radical of Butler's arguments is that what we call "sex" is as much an artificial construct as what we call "gender"; for Butler, therefore, the distinction between the two terms has no validity. Obviously, I am not in complete agreement with Butler on this point, as my readings rely on more (by now) traditional notions which indeed distinguish between nature (sex) and culture (gender). The distinction between the two terms, in the context of such a convoluted plot as that of *Mademoiselle de Maupin*, is challenging to maintain and almost inevitably confusing at times; the challenge and confusion mirror the very ambiguity of the notions as thematized in the text itself.

7. See, among others, René Giraud, "Winckelmann's Part in Gautier's Perception of Classical Beauty," *Yale French Studies* 38 (1967); Georges Poulet, *Trois Essais de mythologie romantique* (Paris: José Corti, 1966); and Weil, "Veiling Desire." See also Gautier's "Du Beau dans l'art," *Revue des deux mondes*, 1 September 1847.

8. *Mademoiselle de Maupin*, 23. These and all subsequent citations from the novel and its preface come from the Adolphe Boschot edition (Paris: Garnier, 1966).

9. The preface is, to a great extent, a Romantic polemic against contemporary conservative and moralistic critics. It is usually identified as the manifesto of the philosophy of "l'art pour l'art." For a contrasting view, see Rosemary Lloyd, "Gautier est-il aussi partisan de la doctrine de l'art pour l'art qu'on veut nous le faire croire?" *Bulletin des etudes parnassiennes* 7 (June 1985): 1–13.

10. See especially 2, 24–25, 37.

11. I am particularly endebted to Sadoff for having found and cited these very suggestive passages from Zola. They are from Zola's *Documents littéraires; études et portraits*, 3d ed. (Paris: Eugène Fasquelle, 1882), 143, 146. See Sadoff, "Ambivalence," 9–10.

12. In her dissertation, "La Femme dans l'oeuvre de Théophile Gautier: de la description à l'interprétation," Nathalie David-Weill makes a suggestive contrast between the uses of description by Balzac and Gautier. Whereas Balzac's goal is primarily to use the physical as an index to the sociological, Gautier's goals are simply esthetic: "les descriptions inondent les textes de Gautier, et son système descriptif ne se situe pas dans un monde symbolique comme chez Balzac, où chaque élément de la description renvoie à un signifié" (13). Carmen Fernandez-Sanchez points out the impossibility of using Gautier's portraits to read characters: "ces portraits, les descriptions des corps et des vêtements, ne servent pas à déchiffrer psychologiquement le personnage. Au contraire, il ne s'agit que d'apparences qui peuvent tromper l'observateur car le dehors de l'individu ne correspond pas à sa véritable nature . . ." See "*Mademoiselle de Maupin* et le récit poétique," *Bulletin de la société Théophile Gautier* 3 (1981): 5. Marie-Claude Schapira, for her part, defines Gautier's descriptive agenda as that of a painter, a purelty esthetic enterprise: "Contrairement à Zola, à Balzac, à Flaubert même, T. Gautier n'a pas besoin d'apprendre à observer. Aucune application dans son talent; son oeil de peintre est bien

éduqué, il a la vision totale spontanée. Les scintillements, les éclairs prismatiques, les reflets humides servent à faire vivre les contours. Mais si l'oeil restitue dans son ensemble le tableau la sensibilité se refuse à l'interpréter." See *Le Regard de Narcisse: romans et nouvelles de Théophile Gautier* (Lyon: Presses Universitaires de Lyon/Editions du CNRS, 1984), 78.

13. See my Chapter 3 on Lavater.

14. See Bouchard, who characterizes Gautier in the following terms: "un écrivain voué au culte de la Beauté, et passionnément attentif, de par son avidité et son acuité visuelles, à la livrée extérieure des choses et des êtres, au point de vouloir faire, selon ses propres termes, d'une 'belle forme une belle idée,' et de 'l'apparence un absolu'" ("Le Masque et le miroir," 583).

15. Some critics maintain, for historical reasons, that the preface and the novel itself are totally unrelated. See Grant, *Théophile Gautier:* "René Jasinski has shown in detail that the preface was written in response to pressures of government censorship that plagued young writers of the 1830's and that it has practically no connection with the fiction . . . As a consequence, the preface is of little value in clarifying the nature of the story . . ." (35). Grant's interpretation is curious, given his refutation elsewhere of the validity of biographical criticism (see note 2, above). In *Le Regard de Narcisse*, Marie-Claude Schapira gives a similar biographical reading of the preface and characterizes it as "l'expression d'une indignation puissante et juvénile" (174). In her introduction to a recent English translation of the novel, Joanna Richardson asserts that the preface "bears little relation to the novel itself, and is probably included on its own brilliant merits as a polemic." See *Mademoiselle de Maupin*, trans. Joanna Richardson (Middlesex, England: Penguin Books, 1981), 18. Cited by Sadoff, "Ambivalence," 183. In contrast, Michel Crouzet emphasizes the continuity between the preface and the fictional narrative: "Bien que nés à part, et de mobile séparés, roman et préface constituent un thème et affirment une seule vérité continue. Le mystère du beau et du corps . . ." "*Mademoiselle de Maupin* ou l'Eros romantique," *Romantisme* 8 (1974): 2.

16. See, among others, Jasinski, *Les Années romantiques,* 313; and Fauchereau, *Théophile Gautier,* 54–55: "un brillant fourre-tout, de structure très imprécise, dans lequel se perdent plus d'une fois le lecteur et l'auteur lui-même qui ne s'en formalise guère . . . Le lecteur se laisse [pourtant] entrainer dans une histoire qui ne tient pas debout."

17. Cited by Jasinski, *Les Années romantiques* 319–20.

18. Fernandez-Sanchez, "*Mademoiselle de Maupin* et le récit poétique," 8.

19. For a discussion of the role of *mal du siècle* in *Mademoiselle de Maupin,* see ibid., 2: "Dans *Mademoiselle de Maupin* le mal du siècle se réduit à sa forme essentielle, à un conflit insoluble entre le réel et l'idéal . . ." For a discussion of the biographical relevance of the *mal du siècle,* see Jasinski, *Les Années romantiques,* 303.

20. The theme of the androgyne in nineteenth-century French literature has been commented on extensively, particularly with regard to Balzac and George Sand. For commentary on Gautier's use of the theme, see Isabelle de Courtivron, "Weak Men and Fatal Women: The Sand Image" in *Homosexuali-*

ties and French Literature: Cultural Contexts/Critical Texts, ed. George Stambolian and Elaine Marks (Ithaca and London: Cornell University Press, 1979), 216; Sadoff, "Ambivalence," and Weil, "Romantic Androgyny." For a literary-historical view of the theme in Gautier, see Joseph Savalle, *Travestis, métamorphoses, dédoublements: essai sur l'oeuvre romanesque de Théophile Gautier* (Paris: Librairie Minard, 1981). Savalle cites Balzac's *La Fille aux yeux d'or* (see my Chap. 4) and Sand's *Lélia* as possible models for *Mademoiselle de Maupin*. Judith Butler characterizes "drag" (transvestism) as a subversive, playful act which points up the artificiality of any notion of "true" gender. See *Gender Trouble*, 136–39. For an expansive overview of transvestism in all its many forms and implications, both historical and contemporary, see Marjorie Garber, *Vested Interests: Cross-Dressing and Cultural Anxiety* (New York and London: Routledge, 1992).

21. Many critics have commented on the relation between the ideal and the real in Gautier. See especially Crouzet, "Mademoiselle de Maupin ou l'Eros romantique"; Smith, *Ideal and Reality*.

22. See Poulet, *Trois Essais de mythologie romantique*, on the obsession shared by Gautier and Nerval for the image of this particular physical type (blond hair, black eyes).

23. See Ross Chambers, "Pour une poétique du vêtement," in *Poétiques* (Ann Arbor: University of Michigan Romance Studies, 1980). Chambers makes the extremely accurate and suggestive observation that certain nineteenth-century poets privileged women's clothing over women's bodies, and points out that "le fétichisme a ceci de particulier qu'il privilége de la femme ce qui n'est pas elle" (19–20). Chambers further explores this phenomenon as an allegory of a "nouvelle conception de l'écriture." Schapira also comments on this fetishism: "Le séduction qu'exerce la femme tient à son corps mais plus encore à sa parure . . . elle dépend surtout de son enchâssement dans un écrin somptueux qui la pare des éclats d'un artifice tout baudelairien . . . une double confusion s'établit entre l'amour et la beauté, entre la beauté et la richesse" (*Le Regard de Narcisse*, 44).

24. Boczenowski explains that d'Albert has "little use for amorphous or veiled figures which might symbolize the virtues of the nonphysical world." She continues by saying that "According to Gautier, the human body found virtue in the very spectacle of its own beauty." See Gautier, *Les Beaux-Arts en Europe* I (Paris: Michel Lévy Frères, 1855), 118: "la forme la plus parfaite que l'homme puisse concevoir est la sienne propre. Son imagination ne saurait aller au delà . . ." (Cited by Boczenowski, "Art and Nature," 118). See Boczenowski, 117–18.

25. There is a possible literary historical explanation for this somewhat anti-intellectual urge: that of Romanticism as the anti-Englightenment. As Jasinski reminds us, "le romantisme naît de deux siècles d'intellectualité, le dix-septième, celui de la raison, et surtout le dixhuitième, celui du raisonnement et de l'esprit." See *Les Années romantiques*, 304.

26. Weil points out the implications of this esthetic system: "If, in the

modern world, art or beauty is only conceived of in feminine terms, this is because its function is ultimately related to and subservient to the artist-lover who is brought to his fullest potential, not in fusion with the woman, but in absorption and mastery of the 'feminine.'" See "Romantic Androgyny," 352. D'Albert's strictly gendered definition of beauty in which men are the appreciative subjects, women the appreciated objects may also be read as an instance of the traditional equation of mind and masculinity, body and femininity. See Butler, *Gender Trouble*, 12.

27. I have chosen to call the title character of the novel "Théodore" when referring to him/her from d'Albert's or Rosette's point of view (this is the only name they know for him/her), "Madelaine" when referring to her own point of view, and "Madelaine/Théodore" from the reader's point of view. Any confusion thus created is, I hope, excusable as it is in keeping with the confusion of the character's identity in the text.

28. Ross Chambers, *La Comédie au château* (Paris: José Corti, 1971), 106–7.

29. Sadoff, "Ambivalence," 139, cites Lucien Dällenbach as attributing the idea of introducing the play within the novel to Nerval. Lloyd suggests that it is the play's lesbian implications that rendered it an appealing (and appropriate) choice for Gautier: "Gautier seems to have found inspiration in *As You Like It* not just for the debate on the theatre and the double *travestissement* of Madelaine: after all, he could have used *Cymbeline* or *Twelfth Night* if he had merely needed a play in which a girl dresses as a boy. *As You Like It*, in addition to its attractive glimpse of the utopian life in the forest of Arden . . . offers a vision of an intimate friendship between two girls . . ." See Lloyd, "Rereading," 28. Marjorie Garber cites *Mademoiselle de Maupin* in the context of her discussion of transvestism in *As You Like It*, and the exploitation of the character of Rosalind by various nineteenth- and twentieth-century authors. See *Vested Interests*, 72–75.

30. See Weil, "Veiling Desire," 219, on d'Albert's need for "firm sexual distinctions." Weil equates this need with a "reliance on the very bourgeois values he claims to despise." See Butler, *Gender Trouble*, 6, on the question of sex and gender. Butler argues that the separation of sex and gender, taken to its most logical extreme, implies that gender escapes the binary categories of sex, and thus "gender itself becomes a free-floating artifice."

31. Weil, "Romantic Androgyny," 355.

32. Weil, "Veiling Desire," 222. See also Butler, *Gender Trouble*, 6–7.

33. The comparison with Laclos's treatment of the same theme is obvious; indeed, Madelaine could be called a sort of "anti-Cécile." For a comparison of *Mademoiselle de Maupin* and *Les Liaisons dangereuses*, see A. and Y. Delmas, *A la Recherche des "Liaisons dangereuses"* (Paris: Mercure de France, 1964).

34. Bouchard, "Le Masque et le miroir," 592–93. At another moment, in reference to the play within the novel, Bouchard notes: "Ici encore, le masque a fonctionné comme miroir . . ." (599).

35. See Sadoff, "Ambivalence," 136: "Lies on Madelaine's part lead to truth on [men's] part—and on hers. Not only does she get to know men inside out,

so to speak . . . but she also discovers her true "self" in the process. This is not totally unexpected, for, as Gaston Bachelard notes, masks can both conceal and reveal various latent parts of one's personality . . ."

36. Critics have been uncharacteristically circumspect about the issue of homosexuality in *Mademoiselle de Maupin*. It seems to me to be of central import to the novel. One exception is Isabelle de Courtivron, who states clearly and intelligently the presence of the theme: "The switching to clothes of the opposite sex is in fact an initiation rite which allows her gradually to lose the consciousness of her sexual identity . . . What she finds on the other side is the acceptance and concretization of her true androgynous impulse, that is, of her bisexuality . . . Rosette, unaware that Théodore/Madeleine is not a man, makes such provocative advances that Madeleine's desire is aroused and she cannot refrain from responding" ("Weak Men and Fatal Women," 217). Courtivron also points out an essential element of the treatment of homosexuality in the novel, the fact that male homosexual desire (far more threatening a prospect) is merely suggested in order to be replaced by a "safer" lesbian scenario: "D'Albert is also androgynous, but only potentially . . . Yet, owing to the stronger social, psychological, and personal taboo of male homosexuality, Gautier can go no further than to project the homo- and bisexual fulfillment onto the female character—or the female self of his hero's persona. In leading her to the furthest stage of sexual androgyny, Gautier permits the acting out of his hero's forbidden and repressed longings, while managing to avoid what would prove a threatening self-confrontation." Courtivron also places the theme in its literary-historical context, noting that Madelaine is more "authentically androgynous" than most of her counterparts in nineteenth-century literature ("which usually tended to abstract versions of this theme—statues, angels, or asexual creatures like Balzac's Séraphita"). Perhaps most suggestively, she points out that androgyny and homosexuality are always, to some degree, linked: "Certainly, the links between androgyny, homosexuality, bisexuality, and transsexuality were not precisely defined in the nineteenth century; they are difficult to distinguish even given our present state of psychoanalytic knowledge. It would therefore be dangerous to categorize any literary character as explicitly homosexual or transsexual. Yet it is also true that from our perspective we can see in much of this earlier literature the dramatization of desires, the imaginative acting out of fantasies which, although impossible to delineate clearly, are equally impossible to ignore" (216). Joseph Savalle sees both d'Albert and Madelaine as repressed homosexuals, with the difference being that Madelaine overcomes her repression, while d'Albert remains a frightened "puritan." See Savalle, *Travestis*. Rosemary Lloyd's suggestion that Gautier chose to include *As You Like It* in his novel specifically for its lesbian implications would underscore the importance of the theme of homosexuality in the text. See my note above. Unfortunately, other critics have been less clear-sighted and have insisted on dismissing, if not ignoring entirely, the homosexual element of the novel. For example, Crouzet ("Mademoiselle de Maupin ou l'Eros romantique") explains that the novel reduces the story of the 17th-century transvestite actress Maupin to a single "aventure, dont

au reste l'"immoralité" est méthodiquement adoucie; quelques coups d'épée, des sauts de barrière, un enlèvement bien vague, un déguisement qui mène à quelques caresses entre filles . . ." (5). Crouzet, it must be said, is willing to accord Madelaine status as a "femme virilisée, membre d'un troisième sexe" (8), or as a mythical figure of *dédoublement;* however, he is apparently unwilling to see, even in the ultimate consummation of the desire between Madelaine and Rosette that he acknowledges, the very real theme of homosexuality in the text as representing anything more than "quelques caresses entre filles." Particularly in light of Courtivron's remarks above, one can only wonder if Crouzet would have found "quelques caresses" between two boys quite as easy to dismiss. Other critics have read the entire theme of the androgyne as a veiled expression of homosexual tendencies on Gautier's part. See Pierre Albouy, "Le Mythe de l'androgyne dans *Mademoiselle de Maupin,"* in *Mythographies* (Paris: José Corti, 1976), 601: "Roman des angoisses de l'artiste et de son impuissance, *Mademoiselle de Maupin* renfermerait-elle une autre confidence de Gautier sur une autre sorte de difficulté—celle de l'amour condamné, de l'homosexualité? La question se pose d'autant plus qu'il existe, chez Gautier, une sorte de hantise de l'hermaphrodite . . ." See also Jean Pierrot, *L'Imaginaire décadent 1800–1900* (Paris: PUF, 177), 168. Cited by Sadoff, "Ambivalence," 166.

37. Sadoff emphasizes d'Albert's need to eliminate all ambiguity. See "Ambivalence," 159. Indeed, Sadoff's entire study focuses on ambiguity as a fundamental principle of Gautier's text. For commentary on the "natural" in the domains of sex and sexuality as a discursive construct, see Butler, *Gender Trouble.*

38. This false conclusion is taken at face value by most critics. For example, Anne Bouchard tells us: "La clé de cette énigme restera pourtant encore en attente, et c'est sur cette attente qu'est batie toute la suite du roman. Par un double jeu fort habile, Gautier, tout en mettent le lecteur dans le prétendu secret, l'oblige à cheminer avec d'Albert, à suivre ses fausses pistes . . . ses hésitations . . . ses conjectures . . . jusqu'à l'irréfutable révélation." While I cannot argue that the love scene between d'Albert and Madelaine does not represent an irrefutable revelation of anatomical gender, I would argue that there is a more profound question at stake here than d'Albert (and most critics) is willing to recognize. See Bouchard, "Le Masque et le miroir," 595.

39. As Weil tells us: "Madelaine/Théodore poses the question to d'Albert of how sexuality can be thought outside of the polarities of gender." See "Veiling Desire," 220. Similar ideas are germane to Butler's project as well.

40. Chambers, *La Comédre au château,* 107–8.

41. Lloyd, "Speculum Amantis, Speculum Artis," 82–83.

42. Weil, "Romantic Androgyny," 357–58.

43. Weil, "Romantic Androgyny," 358; Weil's page references are to the Garnier-Flammarion edition, ed. van den Bogaert (Paris, 1966).

44. Indeed, *Mademoiselle de Maupin* has been characterized as an apology for homosexuality. See A. E. Carter, *The Idea of Decadence in French Literature, 1830–1900* (Toronto: University of Toronto Press, 1958), 39–42. Cited by Boczenowski, "Art and Nature," 26. In her study, Sadoff cites several critics

from various periods who have characterized the novel as something akin to a "monument to immorality." This reputation is due no doubt to its homosexual content, as many critics must have seen in it the "apologia for lesbian love" that Mario Praz did in 1951 (an interpretation which, if neutral politically, would not be entirely inaccurate, but whose hostility I object to). Sadoff also informs us that the novel was put on trial in the state of New York in 1928 in the case of *Halsey versus the New York Society for the Suppression of Vice*. See Sadoff, "Ambivalence," 2–4.

45. Savalle accurately characterizes d'Albert as a latent puritan: "Certes, il a trouvé enfin l'incarnation de son rêve de beauté parfaite, il a, comme il le dit 'trouvé corps à [son] fantôme,' mais les remords et les craintes qui sont les siennes quand il évoque 'la griffe de Satan' semblent lui interdire l'accès à cette beauté maintenant qu'il l'a trouvée; d'où son désarroi et son effroi, dûs essentiellement à son puritanisme." See *Travestis*, 44.

46. I use the word "moral" in quotation marks here in order to distance myself from the antiquated and objectionable view that the "choice" of one's sexuality is a moral question, a view which inevitably comes down to the very simple formula defining heterosexuality as moral, homosexuality as immoral.

47. My emphasis on the contrast between the two characters is in contradiction to many critics who have insisted on readings which identify Madelaine and d'Albert as doubles of each other. See, among others, Savalle, *Travestis*, 50: "Maupin et d'Albert apparaissent être deux formes du même personnage, deux incarnations de leur créateur . . . les deux personnages poussent la gémellité jusqu'à être interchangeables." See also Fernandez-Sanchez, "*Mademoiselle de Maupin* et le récit poétique," 6.

48. Not all critics of the novel would agree with this reading. See Sadoff, "Ambivalence," 161–63, on the debate over Mademoiselle de Maupin's feminist implications.

49. Lloyd, "Rereading," 28.

50. See Sadoff, "Ambivalence," 156: "Gautier's far-sighted heroine deconstructs the notion of the indissoluble link of sex and gender . . ."

51. I would not, of course, go so far as to call Gautier a feminist. Much of his work, like that of all of his (male) contemporaries, is filled with content that can easily be characterized as misogynist. It is important to note, however, that the explicitly misogynist content of *Mademoiselle de Maupin* is voiced by d'Albert, a character whose credibility is continually put into question. See Lloyd, "Rereading," 23. Unfortunately, as I have shown above, many critics succumb to the temptation to identify d'Albert and Gautier as one and the same.

Chapter Six. Zola and *le signe de la femme fatale*

1. E. P. Gauthier, "New Light on Zola and Physiognomy," *PMLA* 75 (1960): 297.

2. Prosper Lucas, *Traité philosophique et physiologique de l'hérédité naturelle*

dans les états de santé et de maladie du système nerveux . . . (Paris: J. B. Baillière, 1850).

3. Gauthier, "New Light on Zola," 304.

4. Zola may also have ben influenced by the anthropometric approach to criminology, which defined criminality as a form of congenital atavism. This particular current of thought was more or less contemporaneous with the writing of *Nana*. The best-known proponents of the trend are Alphonse Bertillon (1853–1914), originator of the Bertillon system (or "bertillonnage") of identifying criminals by anthropometric indices (to which fingerprints were merely a later supplement) and Cesare Lombroso (1835–1909), who developed the concept of the atavistic criminal. See Jeffrey Mehlman, "Craniometry and Criticism," *Boundary 2: A Journal of Postmodern Literature and Culture* 11, 1–2 (Fall–Winter 1982–83): 81–101, and Richard Kellogg, "Lombroso and the Born Criminal," *Baker Street Journal: An Irregular Quarterly of Sherlockiana* 35, 3 (Sept. 1985): 143–45.

5. I learned that the two works were first published simultaneously in the same journal from reading Charles Bernheimer, *Figures of Ill Repute: Representing Prostitution in Nineteenth-Century France* (Cambridge: Harvard University Press, 1989), 214.

6. Zola, *Le Roman expérimental* (Paris: Garnier-Flammarion, 1971), 59. All further references are to this edition.

7. For excellent documentation on Zola's preparations for his novels, see the *Etude* devoted to each by Henri Mitterand in his Pléiade edition of the *Rougon-Macquart* (Paris: Gallimard, 1960–67). See also Mitterand's recent edition of Zola's previously unpublished *Carnet d'enquêtes* (Paris: Plon, 1986). This record shows that Zola's haphazard efforts at sociological fieldwork, while of great anecdotal interest, can hardly be deemed "scientific" investigation.

8. In *Le Réalisme selon Zola: archéologie d'une intelligence* (Paris: PUF, 1975), Alain de Lattre gives a very clear account of the insurmountable contradiction inherent in Zola's theory: "Dès le moment que l'on varie sur notre fantaisie, que l'imagination seule défait, retient, observe pour conclure, plus rien ne se construit, plus rien ne se produit, on brode sur des songeries. La prétention de faire ici de l'expérience, ou d'expérimenter, va contre l'idée même de réalité qu'il s'agit d'établir. Un réalisme expérimental, dans le roman qui est devenue oeuvre d'imagination, c'est une contradiction dans les termes" (79). Unfortunately, de Lattre becomes much less clear and much less convincing when he tries to go "plus loin" than this analysis by telling us that the laboratory is a privileged space, as is the imagination of the novelist. He tries to establish that in fact the analogy between science and fiction can be said to work, but seems to fall prey to Zola's rhetoric by forgetting his original premise that experimentation and imagination are simply and inescapably antithetical. De Lattre's is a bold, if unsuccessful, attempt to go against the critical commonplace that *Le Roman expérimental* is itself a work more of fantasy than of theory. The tradition of this reading began with Ferdinand Brunetière's critique in the *Revue des deux mondes* (15 February 1880) immediately following the publication of Zola's treatise.

9. Gillian Beer, "Plot and the Analogy with Science in Later Nineteenth-Century Novelists," *Comparative Criticism* 2 (1980): 135–36, 137.

10. See Brooks, *Reading for the Plot.*

11. Beer makes the significant observation that "Evolutionary theory is essentially narrative in form: it works through time, and time is its essential condition" ("Plot and the Analogy with Science," 135).

12. The criticism on Zola is dominated by studies, many of them more or less psychoanalytical in perspective, on the "myths" inherent in Zola's work. The best and most often cited of these is Jean Borie's *Zola et les mythes, ou de la nausée au salut* (Paris: Klincksieck, 1971).

13. Among the many critics who have written about Nana as femme fatale are Naomi Schor, Jean Borie, and Chantal Bertrand-Jennings. Their readings of the ideological implications of this myth as exemplified by Zola's novel will be included below in a discussion of the topic. On the image and myth of the femme fatale in late-nineteenth-century art and literature (including *Nana*), see Bram Dijkstra's *Idols of Perversity: Fantasies of Feminine Evil in Fin-de-Siècle Culture* (New York and Oxford: Oxford University Press, 1986).

14. In "The Metaphorical Web in Zola's *Nana*," *University of Toronto Quarterly* 47, 3 (Spring 1978): 239–58, Peter V. Conroy accurately states this idea as follows: "Although his exact descriptions point the novel toward naturalism, Zola's subjective imagination predominates. Because of these images and the metaphorical web they spin about Nana, this novel belongs to a mythic dimension. Nana is not a 'real' woman; rather she is *par excellence* a literary creation" (256).

15. Indeed, it could be said that the entire *Rougon-Macquart* cycle is a mythologization of science.

16. Zola, *Nana* (Paris: Garnier-Flammarion, 1968), 74. All subsequent references are to this edition.

17. There are, of course, women in the audience as well. However, as Naomi Schor convincingly points out in *Zola's Crowds* (Baltimore: Johns Hopkins University Press, 1978), 89–91, the audience is definitely divided into two "crowds" according to gender; it is clearly the reaction of the male "crowd" to Nana's body that is Zola's concern.

18. In "Narrative Tension in the Representation of Women in Zola's *L'Assommoir* and *Nana*," *L'Esprit créateur* 25, 4 (Winter 1985): 93–104, Katherine Slott reads this description of the first appearance of Nana, as well as subsequent ones, as being filtered through male "focalizer" characters in the novel (101–2). For an analysis of color in the novel (which, incidentally, identifies pink and gold as the most prevalent and significant colors), see Alain Pagès, "Rouge, jaune, vert, bleu: etude du système des couleurs dans *Nana*," *Cahiers naturalistes* 49 (1975): 125–35.

19. For a discussion of the male gaze in *Nana*, as seen through a psychoanalytic framework, see Peter Brooks, "Storied Bodies, or Nana at Last Unveil'd," *Critical Inquiry* 16, 1 (Autumn 1989): 1–32.

20. Quoted by Henri Mitterand in his *Etude* of *Nana*, vol. 2 (1961) of his Pléiade edition, 1670 and 1677.

21. Janet Beizer, "Uncovering Nana: The Courtesan's New Clothes," *L'Esprit créateur* 25, 2 (Summer 1985): 47. See also Beizer, "The Body in Question: Anatomy, Textuality, and Fetishism in Zola," *L'Esprit créateur* 29, 1 (Spring 1989): 50–60.

22. This idea is very much in keeping with some of Zola's rather puritanical attitudes toward sex. See Jean Borie's *Zola et les mythes* for commentary on Zola and repression of the body.

23. It is interesting to note that the only men whom Nana seems to trust are those who are immune to her sensual attraction: Francis, the hairdresser, and Labordette, "l'ami des femmes." Both characters are obviously meant to be homosexual, although Zola, in a curious gesture of *pudeur*, never makes this explicit.

24. The idea of Nana as *bête* is already made quite explicit in Zola's *ébauche* for the novel, in which he describes the "philosophical subject" of the novel as that of a female dog chased by a pack of dogs ("une meute derrière une chienne . . ."). See Conroy, "Metaphorical Web," 240, on the *bête* metaphor and its obsessive recurrence in *Nana*. See also Brian Nelson, *Zola and the Bourgeoisie: A Study of Themes and Techniques in the Rougon-Macquart* (Totowa, N.J.: Barnes and Noble, 1983): 12. For a discussion of the woman-as-animal theme frequently found in late-nineteenth-century male fantasies of feminine evil as evinced in both literature and the plastic arts, see Dijkstra, *Idols of Perversity,* 275–332.

25. Chantal Bertrand-Jennings, *L'Eros et la femme chez Zola* (Paris: Klincksieck, 1977), 59.

26. See Brooks, "Storied Bodies," on Nana's nudity as merely the illusion of an unveiling, the removal of a veil which merely reveals (to the masculine gaze) another veil. See also Brooks, "The Body in the Field of Vision," for a discussion of the equivalence, in the context of psychoanalysis, of the erotic components of the Freudian definitions of notions of the desire to see ("scopophilia") and the desire to know ("epistemophilia"). In "The Body in Question," Beizer also discusses these notions and their importance to Zola's work.

27. For a provocative discussion of the denuding of Nana and its significance to the text as a whole, see Brooks, "Storied Bodies."

28. Naomi Schor notes the very significant fact that the disfigured Nana is seen only by women and not by the men, who wait downstairs in the hotel lobby, in her reading of the "double crowd structure" and its definition of sexual difference in *Nana*. See *Zola's Crowds*, 103–4.

29. Beizer, "Uncovering Nana," 56.

30. See Bernheimer, *Figures of Ill Repute,* 213–25, for a complex reading of Nana's death and decomposition which puts into question, in part, the reading to which I subscribe, by which Nana's rotting body reveals her essential self. In the fascinating "Black Bodies, White Bodies: Toward an Iconography of Female Sexuality in Late Nineteenth-Century Art, Medicine, and Literature," *Critical Inquiry* 12, 1 (Autumn 1985): 204–42, Sander Gilman tells us that what is revealed by the dissolution of Nana's mask is specifically the horror of sexuality: "Nana's childlike face is but a mask which conceals the hidden disease

within, the corruption of sexuality. Thus Zola concludes the novel by revealing the horror beneath the mask: Nana dies of the pox. (Zola's pun works in
French as well as in English and is needed because of the rapidity of decay
demanded by the moral implication of Zola's portrait. It would not do to have
Nana die slowly over thirty years of tertiary syphilis. Smallpox, with its play
on 'the pox,' works quickly and gives the same visual icon of decay.) . . . The
decaying visage is the visible sign of the diseased genitalia through which the
sexualized female corrupts an entire nation of warriors and leads them to
the collapse of the French Army and the resultant German victory at Sedan.
The image is an old one, it is *Frau Welt,* Madam World, who masks her
corruption, the disease of being a woman, through her beauty" (235). Bernheimer also notes (224) the suggestive semantic rapprochement to be made
between the terms *vérole* and *petite vérole.* See also, on Nana's death and decay, Bertrand-Jennings, *L'Eros et la femme,* 71, and Jill Warren, "Zola's View
of Prostitution in *Nana,*" in Pierre L. Horn and Mary Beth Pringle, eds., *The
Image of the Prostitute in Modern Literature* (New York: Ungar, 1984), 37. Helena
Michie discusses disfigurement by syphilis as one means of "the erasure of the
prostitute's body" in British literature of the mid-nineteenth century, and emphasizes that the prostitute's "invisibility" suggests "her function as a symbol
or figure rather than human being" (*Flesh Made Word,* 71). Michie also argues
that the Victorian novels she studies use their heroine's bodies as figures of
larger social or criminal issues, as answers to the questions the texts pose concerning these matters (119). I have of course made similar arguments about
the character of Nana.

31. There is an interesting and unmistakable intertextual relation between
Nana's death and disfigurement and the smallpox contracted by the Marquise
de Merteuil at the end of Laclos's *Les Liaisons dangereuses:* "Le sort de Mme. de
Merteuil paraît enfin rempli, ma chère et digne amie . . . Elle en est revenue
[de la petite vérole], il est vrai, mais affreusement défigurée . . . Le Marquis
de *** . . . disait hier, en parlant d'elle, que la maladie l'avait retournée, et qu'à
présent son âme était sur sa figure. Malheureusement tout le monde trouva
que l'expression était juste" (Lettre 175).

32. See Brooks, "Storied Bodies," 18: "One notes [in the "Mouche d'or"
article] the equation, typical in Zola, between a strong female sexuality and
the lower classes: the body as a source of class confusion, of potential revolution, as an object of fear." Also on the "Mouche d'or" article, see Dijkstra,
Idols of Perversity, 240.

33. It is no coincidence that the most authoritative treatise of the period on
the subject of prostitution was written by a public health official, a "membre
du Conseil de salubrité de la ville de Paris." A. J. B. Parent-Duchâtelet's two-
volume *De la prostitution dans la ville de Paris, considérée sous le rapport de l'hygiène
publique, de la morale et de l'administration* (Paris: J. B. Baillière, 1857) gives a
detailed record, based to a large extent on police statistics and interviews with
prostitutes, of the various aspects of the lives of prostitutes. One wonders if
Zola was familiar with the work, as it includes information which seems to
parallel some of his profile of Nana. For example, in chapter 2 of vol. 1,

entitled "Moeurs et habitudes des prostituées," Parent-Duchâtelet gives us the following view of the familial origins of most prostitutes: "L'inconduite des parents et les mauvais exemples de toute espèce qu'ils donnent à leurs enfants doivent être considérés pour beaucoup de filles, et en particulier à Paris, comme une des causes premières de leur détermination. Les dossiers de chaque fille et les procès-verbaux des interrogatoires font sans cesse mention de désordre dans les ménages . . . Ainsi la dépravation, l'insouciance, la position nécessiteuse de beaucoup de gens de la dernière classe, provoquent, ne préviennent pas ou ne peuvent empêcher la corruption des enfants; on peut dire en général pour un bon nombre de prostituées ce que l'observation de tous les jours apprend à l'égard des malfaiteurs, c'est qu'ils ont pour la plupart une origine ignoble" (102). Parent-Duchâtelet, incidentally, also gives some idea of the general physiognomical traits of the prostitute. The following sentence clearly makes one think of Nana: "L'embonpoint de beaucoup de prostituées et leur brillant état de santé frappent tous ceux qui les regardent en masse et qui les voient réunies en assez grand nombre" (1: 185). See Bernheimer's chapter on Parent-Duchâtelet, "Parent-Duchâtelet: Engineer of Abjection" (*Figures of Ill Repute*, 8–32), for a very pertinent reading of Parent's text as indicative of some of the dominant (male) fantasms of its period, much as *Nana* is. For further commentary on late-nineteenth-century theories and imagery of prostitution, see Dijkstra, *Idols of Perversity*, 352–75.

34. Philippe Hamon, "Zola: romancier de la transparence," *Europe* 46, 468–69 (April–May 1968): 390.

35. Although she uses the concept to different interpretative ends, as we saw above, the concept of corporality as substitute for textuality is one of Beizer's central theses.

36. At least one critic interprets the function of the "Mouche d'or" episode in the novel in a completely contrary fashion. Per Buvik, in his "Nana et les hommes," *Cahiers naturalistes* 49 (1974): 105–24, maintains that the "Mouche d'or" and indeed all of *Nana* is merely intended as parody of the bourgeois myth of the femme fatale: "Le portrait qu'il [Fauchery] fait de Nana et de ses ravages est tellement fantastique qu'il est proche d'une caricature des idées puritaines sur la femme" (122). I would maintain, however, that nowhere does Zola display any irony toward such myths, and indeed his entire narrative propagates them. To late-twentieth-century readers, these ideas are indeed caricatural; Buvik, nonetheless, does not succeed in convincing us that Zola's text does not incorporate them in all earnestness.

37. Bernard's most important work is of course the *Introduction à l'étude de la médecine expérimentale* (Paris, 1865) on which Zola's theory of the *roman expérimental* is based. Taine's concept of *la race, le milieu, et le moment*, clearly an influence on Zola, is introduced in his *Histoire de la littérature anglaise* (Paris, 1863). Lucas, the best-known propagator of theories of heredity of his time, is the author of *Traité philosophique et physiologique de l'hérédité naturelle dans les états de santé et de maladie du système nerveux* (cited above). Bernheimer points out that Michelet's writings on women, *L'Amour* (1858) and *La Femme* (1859), also influenced Zola, and that Michelet's ideas on hysteria had been in turn

influenced by Prosper Lucas. See *Figures of Ill Repute*, 202–8. On hysteria as part of late-nineteenth-century misogyny and gynophobia, see Dijkstra, *Idols of Perversity*, 243–44.

38. See Yves Malinas, *Zola et les hérédités imaginaires* (Paris: Expansion Scientifique Française, 1985), on the actual scientific implications of Zola's theories of heredity: "Zola a choisi comme fil conducteur ce qu'on appelait à son époque l' 'hérédité.' Non pas la transmission de caractères morphologiques bien définis mais l'hérédité dans sa forme la plus obscure, la plus difficile à débrouiller: l'hérédité des affections neuro-psychiatriques" (32). Malinas goes on to tell us that Zola was well aware of the uncertainty surrounding his subject: "Zola a conscience de l'instabilité du terrain où il s'engage . . . Il montre la complexité des faits, la progression des hypothèses vers une synthèse toujours incertaine, toujours remise en question" (32). See also Michel Butor, "Emile Zola, romancier expérimental, et la flamme bleue," *Critique* 239 (1967): 407–37. It is revealing to note that this same issue, the genetic transmission of traits without obvious physiological bases (psycho-intellectual and even moral qualities), remains, more than a century later, a controversial topic among geneticists. It is indeed a question which invites endless speculation and fascination among scientists and laymen alike.

39. As far as feminist questions are concerned, Chantal Jennings has written a two-part article specifically addressing the contradiction between Zola's stated opinions and the "message" of his works of fiction: "Zola féministe?" *Cahiers naturalistes* 44 (1972): 172–87; 45 (1973): 1–22.

40. Nana's child's "bad blood" becomes overdetermined when it is suggested, in at least one place, that he may be the result of an incestuous relationship between Nana and a cousin. See Schor, *Zola's Crowds*, 95–96: "Nana's apparently innocent fantasy about a cousin leads us to a peculiar disparity between information provided by *Nana* and by the Genealogical Tree with reference to the father of Nana's child: whereas in the novel, Louiset's father is unknown, evasively described as a 'gentleman,' the family tree specifies that Nana 'has a child by a cousin.' The unfortunate consequence of Nana's family romance—her sickly offspring—is typical of all such relationships featured in Zola's novels . . ."

41. Obviously, I am using the term "vice" as it was used in the nineteenth century. Much of what was then defined as "vice" we would now more accurately define in terms of mental or emotional pathology, as sickness (including alcoholism and extreme promiscuity, with or without monetary compensation).

42. The ideological implications of Zola's society-as-body metaphor and the essentialism and determinism inherent in a view such as Zola's is outlined by Georg Lukács, in his famous *Balzac et le réalisme français*: "Ce que Zola admet . . . comme résultat 'scientifique,' c'est la conception non dialectique d'une unité organique dans la nature et la société, c'est-à-dire l'élimination des contradictions comme fondement du mouvement de la société, une doctrine 'harmonieuse' de l'essence de la société, qui enferme la critique sociale subjectivement honnête et courageuse de Zola dans le cercle magique d'une

étroitesse d'esprit progressiste mais bourgeois impossible à dépasser." Cited by Jean Borie in "Les Fatalités du corps dans *Les Rougon-Macquart*," *Les Temps modernes* 273 (March 1969): 1572, note 5.

43. Bernheimer notes that the "disease" transmitted by Nana is harmful only to those of a different (higher) social class. Nana does not infect or destroy her own kind.

44. Bénédict-Auguste Morel, *Traité des dégénérescences physiques, intellectuelles et morales de l'espèce humaine, et des causes qui produisent ces variétés maladives* (Paris: J. B. Baillière, 1857), 50–51, 461. These passages from Morel's treatise are cited by Jean Borie in *Mythologies de l'hérédité au XIXe siècle* (Paris: Editions Galilée, 1981), 116. The term and indeed the concept of *classes dangereuses* to which Morel refers seem to have been introduced by H. A. Frégier in his *Des Classes dangereuses de la population dans les grandes villes* (Paris, 1840).

45. Borie, *Mythologies de l'hérédité*, 26, emphasis mine. It is relevant, if not central, to my concerns to read the continuation of Borie's analysis, in which he articulates the contradiction between Zola's political views and his novelistic practice: "Zola, en même temps qu'il le charge, veut absoudre le peuple . . . Si [donc] le peuple en est réduit au corps, c'est qu'il est condamné par la bourgeoisie à rester séparé, à rester peuple . . . Nous pouvons ainsi formuler cette contradiction: le mal vient du peuple, et il est imposé au peuple. La 'générosité' de Zola donne l'absolution, mais la vision social reste bloquée, et le restera aussi longtemps que le corps demeurera prisonnier dans son enfer. Il ne s'agit pas ici évidemment de la *pensée* sociale de Zola . . . mais plutôt, par exemple, de la possibilité pour lui de peindre un homme du peuple qui soit un *homme véritable*, ni brute, ni martyr . . ." (26–27).

46. This notion, in the Judeo-Christian tradition, dates back to the Adam and Eve story, and is certainly one of the most enduring and harmful myths of Western cultures. For a brilliant analysis of the history of women, sexuality, and religion (particularly, but not exclusively, Christianity), see Marina Warner, *Alone of All Her Sex: The Myth and Cult of the Virgin Mary* (New York: Vintage Books, 1983).

47. We must also note, however, the crucial fact that Zola does not let Nana get away with having corrupted the bourgeoisie. Social order is restored through Nana's punishment by death. Chantal Bertrand-Jennings analyzes the ideological implications of Nana's death in "Lecture idéologique de *Nana*," *Mosaic* 10, 4 (1976–77): "Nana sera bien entendue châtiée de tous ses crimes par le dénouement que l'on sait. Il faut cependant remarquer l'ambiguité de cette entreprise de liquidation systématique des refoulés et opprimés de la société bourgeoise que sont le sexe, le peuple et l'insoumise, par l'intermédiaire du personnage de la courtisane. Bouc émissaire chargé de tous les maux et de tous les péchés du monde, Nana est bien la victime sacrificielle dont l'atroce mort expiatoire et conjuratoire vient rétablir l'ordre de la société qui l'a immolée. Et, en dépit de ses velléités naturalistes, le roman dans son ensemble constitue une féroce mise à mort des indésirables de cette même société bourgeoise" (54). On Zola and the two *refoulés* that are the body and the underclass, see (again) Jean Borie. On Zola's sociopolitical agenda, see Brian Nelson,

Zola and the Bourgeoisie. See also Françoise Naudin-Patriat, *Ténèbres et lumière de l'argent: la représentation de l'ordre social dans Les Rougon-Macquart* (Dijon: Université de Dijon, 1981).

48. Among the best and best-known of feminist readings of Zola's work, and particularly of *Nana*, are Chantal Bertrand-Jennings' *L'Eros et la femme chez Zola*, as well as her numerous articles; Naomi Schor's "Le Sourire du sphinx: Zola et l'énigme de la féminité," *Romantisme* 13–14 (1976): 183–95, and "Mother's Day: Zola's Women," *Diacritics* 5, 4 (Winter 1975): 11–17, as well as her *Zola's Crowds*; Katharine Slott's "Narrative Tension in the Representation of Women in Zola's *L'Assommoir* and *Nana*"; Sander Gilman's "Black Bodies, White Bodies"; and Anna Krakowski's *La Condition de la femme dans l'œuvre d'Emile Zola* (Paris: A. G. Nizet, 1974). All of these critics seem to agree that *Nana* exemplifies what Bram Dijkstra has called "the turn-of-the-century male's fascination for, horror of, and hostility toward woman, culminating in an often uncontrollable urge to destroy her . . ." See *Idols of Perversity*, 149.

49. For a discussion of the genetic *fêlure* of the Rougon-Macquart family, and its origin in Tante Dide, see Beizer, "The Body in Question," 52. On nymphomania as a theme of nineteenth-century misogynist thought, see Dijkstra, *Idols of Perversity*, 250.

50. See Bertrand-Jennings, *L'Eros et la femme chez Zola*, 50–54, on "le désir hystérique." As I have noted above, Bernheimer has shown that Zola's ideas on hysteria were influenced by Prosper Lucas, through Michelet.

51. Morel, *Traité*, 324. Both this and the passage from Morel below are cited by Jean Borie in *Mythologies de l'hérédité*, 108–9.

52. Morel, *Traité*, 565, emphasis mine.

53. The *Petit Robert* does even include the dermatological usage in its six subdefinitions of the word *signe*.

54. Bernheimer, *Figures of Ill Repute*, 218, also notes the significance of the mole, and remarks that Lombroso and Ferrero signal an unusually high occurrence of hairy moles among "degenerate" females in their turn-of-the-century studies of criminality.

55. Cited by Henri Mitterand in his *Etude* of *Nana*, 1677.

56. Borie, *Zola et les mythes*, 49.

57. In "Storied Bodies," Brooks convincingly argues that Nana's sex is "the *puissance motrice* of the text." He notes that it is also a *"puissance occulte."* Brooks places *Nana* in its literary-historical context, part of a "larger cultural story," as one of several "late romantic and naturalist narratives, . . . [which] often give a new presence to the body and make it problematic in new ways" (27–28). The texts to which Brooks compares *Nana* are those of the Goncourts, Huysmans, and Barbey d'Aurevilly. Bernheimer takes issue with Brooks's characterization of Nana's sex as unknowable (*"puissance occulte"*) and as the driving force of the text. Bernheimer argues that it is not the female body but rather male fantasms of the female body which propel the narrative, and criticizes what he sees as Brooks's unproblematized use of misogynistic Freudian notions (those of castration and fetishism) in "Storied Bodies," criticisms to which

Brooks replies. See Bernheimer, "Response to Peter Brooks," *Critical Inquiry* 17, 4 (Summer 1991): 868–874; Brooks's reply follows.

58. The theme of castration in *Nana* has been commented on most suggestively, and at greatest length, by Jean Borie in *Zola et les mythes*. It is indeed the central issue of Borie's psychoanalytic reading. See also Roland Barthes, "La Mangeuse d'hommes," *Bulletin mensuel du Guilde du Livre* 20 (June 1955): 226–28, and Bertrand-Jennings, *L'Eros et la femme chez Zola*.

59. Zola's atheistic puritanism is but one of many contradictions in his rather confused ideology.

60. Hamon, "Zola: romancier de la transparence." 391.

61. Ibid.

62. This shared textual agenda is of course in direct contradiction to that we saw in the previous chapter with Gautier's *Mademoiselle de Maupin*.

63. The notion of a chasm between theory and practice within theory itself is of course germane to my reading of Lavater—see Chapter 3. I would argue that such a contradiction is the heart of all physiognomical discourse.

Conclusion

1. Jeanne Fahnestock has argued that, for contemporary readers, the codes of physiognomical character description in mid-nineteenth-century British novels were based on recognizable conventions, a system of signification common to both author and reader. While this may well be true for the Victorian novels Fahnestock studies, the novels I have studied here are "unreadable" for reasons other than our historical distance from them. In contrast to their British counterparts, the corporeal signs presented in these novels do not refer to any generally known physiognomical conventions of the period—in fact, as I have discussed with respect to Balzac and Zola, their corporeal clues are often patently obscure and "unconventional." See Fahnestock, "Heroine of Irregular Features." For her part, Helena Michie identifies and analyzes specifically rhetorical codes (namely, synecdoche, cliché, and metaphor) at work in literary descriptions of women's bodies, within the context of her discussion of "the larger codedness of the body, metaphor, the novel and the realist project itself" (*Flesh Made Word*, 86).

2. See Iser, *Act of Reading*.

Bibliography

Abraham, Pierre. *Balzac et la figure humaine: la couleur des cheveux*. Paris: Bulletin de la Société de Morphologie, 1928.

Abraham, Pierre. *Créatures chez Balzac*. Paris: Gallimard, 1931.

Albouy, Pierre. "Le Mythe de l'androgyne dans *Mademoiselle de Maupin*." In *Mythographies*. Paris: José Corti, 1976.

Amossy, Ruth. "Types ou stéréotypes? Les *Physiologies* et la littérature industrielle." *Romantisme* 18, 64 (1989): 113–23.

André, Jacques, ed. *Traité de physiognomonie*. Paris: Les Belles Lettres, 1981.

(N.B.: All citations of articles published in *L'Année balzacienne* are included under a single entry for the publication; I have organized the individual listings according to the year of their publication.)

L'Année balzacienne.

1962: Gendzier, Stephen J. "L'Interprétation de la figure humaine chez Diderot et Balzac," 181–93.

1965: Metadier, Paul. "Balzac, Précurseur de la caractérologie," 309–16.

1970: Delattre, Geneviève. "De *Séraphîta* à *La Fille aux yeux d'or*," 183–226.

1979: Therien, Michel. "Métaphores animales et écriture balzacienne: le portrait et la description," 193–208.

1982: Labouret-Grare, Mireille. "L'Aristocrate balzacienne et sa toilette," 181–93.

1984: Butler, Ronnie. "La Noblesse d'Empire dans *La Comédie humaine*," 163–78.

1985: Dupuis, Danielle. "La Poésie de la toilette féminine chez Balzac," 173–95.

Guichardet, Jeannine. "Athanase Granson, corps tragique," 152–60.

Guise, René. "Balzac et 'Le Charivari' en 1837," 133–54.

Moret, Nicole. "Alençon, ville-corps," 297–305.

1988: Queffélec, Lise. "*La Vieille Fille* ou la science des mythes," 163–77.

Aristotle. *Physiognomics*. In *Complete Works*, edited by Jonathan Barnes, vol. 1. Princeton: Princeton University Press, 1984.

Arland, Marcel. *Marivaux*. Paris: Gallimard, 1950.

Arnaud, Marie-Anne. "*La Vie de Marianne* et *Le Paysan parvenu*: Itinéraire féminin, itinéraire masculin à travers Paris." *Revue d'histoire littéraire de la France* 82, 3 (1982): 392–411.

Baguley, David, ed. *Critical Essays on Emile Zola*. Boston: G. K. Hall, 1986.

Baldensperger, Fernand. *Orientations étrangères chez Balzac*. Paris: H. Champion, 1927.

Baldensperger, Fernand. "Les Théories de Lavater dans la littérature française." In *Etudes d'histoire littéraire*. Paris, 1910; Geneva: Slatkine Reprints, 1970.

Balzac, Honoré de. Avant-Propos. In Vol. 1 of *La Comédie humaine*. Paris: Furne, Durochet, et al., 1842.

Balzac, Honoré de. *La Fille aux yeux d'or*. In *La duchesse de Langeais, suivi de La Fille aux yeux d'or*, edited by Pierre Barbéris. Paris: Le Livre de Poche, 1972.

Balzac, Honoré de. *Théorie de la démarche*. In *Traité de la vie élégante, suivi de La Théorie de la démarche*, edited by Claude Varèze. Paris: Bossard, 1922.

Balzac, Honoré de. *Traité de la vie élégante*. In *Les Parisiens comme ils sont (1830–1846)*, edited by Andre Billy. Geneva: La Palatine, 1947.

Balzac, Honoré de. *La Vieille Fille*. Edited by Pierre-Georges Castex. Paris: Garnier, 1957.

Barbéris, Pierre. *Mythes balzaciens*. Paris: A. Colin, 1972.

Barkan, Leonard. *Nature's Work of Art: The Human Body as Image of the World*. New Haven and London: Yale University Press, 1975.

Barthes, Roland. "La Mangeuse d'hommes." *Bulletin mensuel du Guilde du Livre* 20 (June 1955): 226–28.

Becker, Colette, ed. *Les Critiques de notre temps et Zola*. Paris: Garnier, 1972.

Beer, Gillian. "Plot and the Analogy with Science in Later Nineteenth-Century Novelists." *Comparative Criticism* 2 (1980): 131–49.

Beizer, Janet L. "The Body in Question: Anatomy, Textuality, and Fetishism in Zola." *L'Esprit créateur* 29, 1 (Spring 1989): 50–60.

Beizer, Janet L. "Uncovering Nana: The Courtesan's New Clothes." *L'Esprit créateur* 25, 2 (Summer 1985): 45–56.

Benesch, Rita. *Le Regard de Théophile Gautier*. Zürich: Juris Druck Verlag, 1969.

Bennington, G. P. "Les Machines de l'opéra: le jeu du signe dans *Le Spectateur français* de Marivaux." *French Studies* 36, 2 (1982): 154–70.

Berkeley, George. "The Theory of Visual Language Showing the Immediate Presence and Providence of a Diety Vindicated and Explained." In *The Works of George Berkeley*, edited by T. E. Jessop and A. A. Luce, vol. 1. London: Thomas Nelson and Sons, 1948.

Bernard, Claude. *Introduction à l'étude de la médecine expérimentale*. Paris, 1865.

Bernheimer, Charles. *Figures of Ill Repute: Representing Prostitution in Nineteenth-Century France*. Cambridge: Harvard University Press, 1989.

Berta, Michael. *De l'Androgynie dans les Rougon-Macquart et deux autres études sur Zola*. New York: Peter Lang, 1985.

Birdwhistell, Ray L. "Background Considerations to the Study of the Body as a Medium of 'Expression.' " In *The Body as a Medium of Expression*, edited by Jonathan Benthall and Ted Polhemus. New York: E. P. Dutton, 1975.

Boczenowski, Joann Marie. "Art and Nature in *Mademoiselle de Maupin* and *A Rebours*." Ph.D. Diss., University of Virginia, 1985.

Bonhôte, Nicolas. *Marivaux ou les machines de l'opéra: étude de sociologie de la littérature*. Lausanne: Editions de l'Age de l'Homme, 1974.

Bonnefis, Philippe. "Le Bestiaire d'Emile Zola." *Europe* 468–69 (April–May 1968): 97–109.

Borie, Jean. "Les Fatalités du corps dans *Les Rougon-Macquart*." *Les Temps modernes* 273 (March 1969).

Borie, Jean. *Mythologies de l'hérédité au XIXe siècle*. Paris: Galilée, 1981.

Borie, Jean. *Le Tyran timide: le naturalisme et la femme au XIXe siècle.* Paris: Klincksieck, 1973.

Borie, Jean. *Zola et les mythes, ou de la nausée au salut.* Paris: Klincksieck, 1971.

Bottéro, Jean. "Symptômes, signes, écritures." In *Divination et rationalité.* Paris: Editions du Seuil, 1974.

Bouchard, Anne. "Le Masque et le miroir dans *Mademoiselle de Maupin.*" *Revue d'histoire littéraire de France* 72, 4 (July–Aug. 1972): 583–99.

Brady, Patrick. "A Decade of Zola Studies, 1976–1985." *L'Esprit créateur* 25, 4 (Winter 1985): 3–16.

Brady, Patrick. *Structuralist Perspectives in Criticism of Fiction: Essays on Manon Lescaut and La Vie de Marianne.* Bern: Peter Lang, 1978.

Brooks, G. P., and R. W. Johnson. "Johann Caspar Lavater's *Essays on Physiognomy.*" *Psychological Reports* 46 (1980): 3–20.

Brooks, Peter. "The Body in the Field of Vision." *Paragraph* 14, 1 (March 1991): 46–67.

Brooks, Peter. *Body Work: Objects of Desire in Modern Narrative.* Cambridge, MA, and London: Harvard University Press, 1993.

Brooks, Peter. *The Melodramatic Imagination: Balzac, Henry James, melodrama and the mode of excess.* New Haven: Yale University Press, 1976.

Brooks, Peter. *The Novel of Worldliness.* Princeton: Princeton University Press, 1969.

Brooks, Peter. *Reading for the Plot: Design and Intention in Narrative.* New York: Vintage Books, 1985.

Brooks, Peter. "Storied Bodies, or Nana at Last Unveil'd." *Critical Inquiry* 16, 1 (Autumn 1989): 1–32.

Brooks, Peter. "The Text of the City." *Oppositions* 8 (1977): 7–11.

Busst, A. J. L. "The Image of the Androgyne in the Nineteenth Century." In *Romantic Mythologies,* edited by Ian Fletcher. London: Routledge and Kegan Paul, 1967.

Butler, Judith. *Gender Trouble: Feminism and the Subversion of Identity.* New York: Routledge, 1990.

Butler, R. "Restoration Perspectives in Balzac's *La Vieille Fille.*" *Modern Languages* 57 (1976): 126–31.

Butor, Michel. "Emile Zola, romancier expérimental, et la flamme bleue." *Critique* 239 (1967): 407–37.

Buvik, Per. "Nana et les hommes." *Cahiers naturalistes* 49 (1974): 105–24.

Cabanis, P. J. G. *Rapports du physique et du moral de l'homme.* Paris: Caille et Ravier, 1815.

Chambers, Ross. *La Comédie au château.* Paris: José Corti, 1971.

Chambers, Ross. "Pour une poétique du vêtement." In *Poétiques.* Ann Arbor: University of Michigan Romance Studies, 1980.

Cismaru, Alfred. "The Misogyny of Marivaux." *Lamar Journal of the Humanities* 11, 1 (Spring 1985): 11–17.

Clapham, G. T. "Lavater, Gall, et Baudelaire." *Revue de littérature comparée* 13 (1933): 259–98.

Coates, Carrol Franklin. "Balzac's Physiognomy of Genius." Ph.D. Diss., Yale University, 1964.

Condillac, Etienne Bonnot de. *Essai sur l'origine des connaissances humaines.* Paris: Librairie Armand Colin, 1924.

Conroy, Peter V., Jr. "The Metaphorical Web in Zola's *Nana.*" *University of Toronto Quarterly* 47, 3 (Spring 1978): 239–58.

Coulet, Henri. *Marivaux romancier: essai sur l'esprit et le cœur dans les romans de Marivaux.* Paris: Armand Colin, 1975.

Coulet, Henri, and Michel Gilot. *Marivaux: un humanisme expérimental.* Paris: Larousse, 1973.

Courtine, Jean-Jacques, and Claudine Haroche. *Histoire du visage.* Paris: Editions Rivages, 1988.

Courtivron, Isabelle de. "Weak Men and Fatal Women: The Sand Image." In *Homosexualities and French Literature: Cultural Contexts/Critical Texts,* edited by George Stambolian and Elaine Marks. Ithaca and London: Cornell University Press, 1979.

Coward, David. *Marivaux: La Vie de Marianne and Le Paysan parvenu.* London: Grant and Cutler, 1982.

Creech, James. "Marivaux in the Classical Episteme, or: Suis-je chez moi, Marie?" *L'Esprit créateur* 25, 3 (Fall 1985): 30–41.

Crouzet, Michel. "Gautier et le problème de 'créer.' " *Revue d'histoire littéraire de la France* 72, 4 (July–Aug. 1972): 659–87.

Crouzet, Michel. "*Mademoiselle de Maupin* ou l'Eros romantique." *Romantisme* 8 (1974): 2–21.

Cureau de la Chambre, Marin. *L'Art de connoistre les hommes.* 2 vols. Paris: Chez Jacques d'Allin, 1667.

Cureau de la Chambre, Marin. *Les Charactères des passions.* Paris: Chez P. Ricolet, 1653.

Damisch, Hubert. "L'Alphabet des masques." *Nouvelle Revue de psychanalyse* 21 (Spring 1980): 123–31.

David-Weill, Nathalie. "La Femme dans l'oeuvre de Théophile Gautier: de la description à l'interprétation." Ph.D. Diss., New York University, 1988.

Delmas, A. et Y. *A la recherche des "Liaisons dangereuses."* Paris: Mercure de France, 1964.

Deloffre, Frédéric. "De Marianne à Jacob: les deux sexes du roman chez Marivaux." *L'Information littéraire* 11, 5 (Nov.–Dec. 1959): 185–92.

Deloffre, Frédéric. *Marivaux et le marivaudage.* 2d ed. Paris: Armand Colin, 1967.

Demetz, Peter. "Balzac and the Zoologists: A Concept of the Type." In *The Disciplines of Criticism: Essays in Literary Theory, Interpretation, and History,* edited by Peter Demetz. New Haven: Yale University Press, 1968.

Descartes, René. *Les Passions de l'âme.* Edited by Geneviève Rodis-Lewis. Paris: Vrin, 1966.

Dictionary of the History of Science. Edited by E. J. Browne, W. F. Bynum, and Roy Porter. Princeton: Princeton University Press, 1981.

Dijkstra, Bram. *Idols of Perversity: Fantasies of Feminine Evil in Fin-de-Siècle Culture*. New York and Oxford: Oxford University Press, 1986.

Eigeldinger, Marc. "L'Inscription du théâtre dans l'oeuvre narrative de Gautier." *Romantisme* 12, 38 (1982): 141–50.

Fahnestock, Jeanne. "The Heroine of Irregular Features: Physiognomy and Conventions of Heroine Description." *Victorian Studies* 24, 3 (Spring 1981): 325–50.

Farag, Gamil. *Le Costume et le caractère dans La Comédie humaine*. Cairo: Livres de France, 1974.

Fauchereau, Serge. *Théophile Gautier*. Paris: Editions Denoël, 1972.

Felman, Shoshana. "Rereading Femininity." *Yale French Studies* 62 (1981): 19–44.

La Femme au XIXe siècle: littérature et idéologie. Lyon: Presses Universitaires de Lyon, 1979.

Fernandez, Ramon. *Balzac*. Paris: Stock, 1943.

Fernandez-Sanchez, Carmen. "*Mademoiselle de Maupin* et le récit poétique." *Bulletin de la société Theophile Gautier* 3 (1981): 1–10.

Fess, Gilbert Malcolm. *The Correspondence of Physical and Material Factors with Character in Balzac*. Philadelphia: Publications of the University of Pennsylvania, 1924.

Frappier-Mazur, Lucienne. *L'Expression métaphorique dans La Comédie humaine*. Paris: Klincksieck, 1976.

Garber, Marjorie. *Vested Interests: Cross-Dressing and Cultural Anxiety*. New York and London: Routledge, 1992.

Gauthier, E. P. "New Light on Zola and Physiognomy." *PMLA* 75 (June 1960): 297–308.

Gauthier, Henri. *L'Image de l'homme intérieur chez Balzac*. Geneva: Droz, 1984.

Gautier, Théophile. *Mademoiselle de Maupin*. Edited by Adolphe Boschot. Paris: Garnier, 1966.

Gheude, Michel. "La Vision colorée dans *La Fille aux yeux d'or* de Balzac." *Synthèses* 289–290 (July–Aug. 1970): 44–49.

Gilman, Richard. *Decadence: The Strange Life of an Epithet*. New York: Farrar, Straus, and Giroux, 1975.

Gilman, Sander L. "Black Bodies, White Bodies: Toward an Iconography of Female Sexuality in Late Nineteenth-Century Art, Medicine, and Literature." *Critical Inquiry* 12, 1 (Autumn 1985): 204–42; reprinted in *Race, Writing, and Difference,* edited by Henry Louis Gates, Jr. Chicago: University of Chicago Press, 1986. 223–61.

Giraud, Raymond. "Gautier's Dehumanization of Art." *L'Esprit créateur* 3, 1 (Spring 1963): 3–9.

Graham, John. "Character Description and Meaning in the Romantic Novel." *Studies in Romanticism* 5 (1966): 208–18.

Graham, John. "The Development of the Use of Physiognomy in the Novel." Ph.D. Diss., Johns Hopkins University, 1960.

Graham, John. *Lavater's Essays on Physiognomy: A Study in the History of Ideas*. Bern: Peter Lang, 1979.

Graham, John. "Lavater's Physiognomy: A Checklist." *Papers of the Bibliographical Society of America* 15 (1961): 308.

Graham, John. "Lavater's Physiognomy in England." *Journal of the History of Ideas* 22 (1961): 561–72.

Grant, Richard B. *Théophile Gautier.* New York: Twayne Publishers, 1975.

Grivel, Charles. "L'Histoire dans le visage." In *Les Sujets de l'écriture,* edited by Jean Decottignies. Lille: Presses Universitaires de Lille, 1981.

Guinaudeau, Oliver. "Au Temps du mesmérisme: un fervent adepte du magnétisme animal, le pasteur J. G. Lavater." *Etudes germaniques* 13, 2 (1958): 98–113.

Guinaudeau, Oliver. *Jean-Gaspard Lavater.* Paris: F. Alcan, 1924.

Hamon, Philippe. "Clausules." *Poétique* 24 (1975): 495–526.

Hamon, Philippe. *Introduction à l'analyse du descriptif.* Paris: Hachette, 1981.

Hamon, Philippe. "Texte littéraire et métalangage." *Poétique* 31 (1977): 261–84.

Hamon, Philippe. "Zola: romancier de la transparence." *Europe* 46, 468–69 (Apr.–May 1968): 385–91.

Harari, Josué. "The Pleasures of Science and the Pains of Philosophy: Balzac's Quest for the Absolute." *Yale French Studies* 67 (1984): 135–63.

Heier, Edmund. "Lavater's System of Physiognomy as a Mode of Characterization in Lermontov's Prose." *Arcadia* 6, 3 (1971): 267–82.

Heier, Edmund. "The 'Literary Portrait' as a Device of Characterization." *Neophilologus* 60 (1976): 321–31.

Hoffmann, Paul. "Marivaux féministe." *Travaux de linguistique et de littérature publiés par l'Université de Strasbourg* 15, 1 (1977): 91–100.

Hogarth, William. *The Analysis of Beauty.* London: J. Reeves, 1753.

Imbert, Patrick. *Sémiotique et description balzacienne.* Ottawa: Editions de l'Université de Ottawa, 1978.

Jameson, Fredric. "The Ideology of Form: Partial Systems in *La Vieille Fille.*" *Sub-Stance* 15 (1976): 29–49.

Jasinski, René. *Les Années romantiques de Théophile Gautier.* Paris: Vuibert, 1929.

Jennings, Chantal Bertrand. "Current Trends in Zola Scholarship." *University of Toronto Quarterly* 50, 3 (Spring 1981): 323–29.

Jennings, Chantal Bertrand. *L'Eros et la femme chez Zola: de la chute au paradis retrouvé.* Paris: Klincksieck, 1977.

Jennings, Chantal Bertrand. "Lecture idéologique de *Nana.*" *Mosaic* 10, 4 (1976–77): 47–54.

Jennings, Chantal Bertrand. "Les trois visages de Nana." *French Review* 44 (special issue 2): 117–28.

Jennings, Chantal Bertrand. "Zola féministe?" *Cahiers naturalistes* 44 (1972): 172–87; 45 (1973): 1–22.

Jennings, Chantal Bertrand. "Zola's Women: The Case of a Victorian 'Naturalist.'" *Atlantis: A Women's Studies Journal/Journal d'etudes sur la femme* 10, 1 (Fall 1984): 26–36.

Jost, François. "George Sand et les *Physiognomische Fragmente* de Lavater." *Arcadia* 12 (1977): 65–72.

Jugan, Annick. *Les Variations du récit dans "La Vie de Marianne" de Marivaux.* Paris: Klincksieck, 1979.

Kanes, Martin. *Balzac's Comedy of Words.* Princeton: Princeton University Press, 1979.

Kapp, Volker. "Le Bonheur de l'instant dans *Mademoiselle de Maupin.*" *Les Lettres Romanes* 38, 1–2 (Feb.–May 1984): 77–97.

Kars, Hendrik. *Le Portrait chez Marivaux: étude d'un type de segment textuel.* Amsterdam: Rodopi, 1981.

Kashiwagi, Takao. *La Trilogie des célibataires d'Honoré de Balzac.* Paris: A. G. Nizet, 1983.

Kempf, Roger. *Sur le corps romanesque.* Paris: Seuil, 1968.

King, Donald L. *L'Influence des sciences physiologiques sur la littérature française de 1670 à 1870.* Paris: Les Belles Lettres, 1929.

Krakowski, Anna. *La Condition de la femme dans l'œuvre d'Emile Zola.* A. G. Nizet, 1974.

Kunz, Reinhard. *Johann Caspar Lavaters Physiognomielehre im Urteil von Haller, Zimmermann und anderen zeitgenössichen Arzten.* Zurich: J. Druck, 1970.

Lapp, John. "The Watcher Betrayed and the Fatal Woman: Some Recurring Patterns in Zola." *PMLA* 24 (June 1959): 276–84.

Larat, Jean. "Un Fragment inédit de Charles Nodier, sa 'Physionomie' inspirée de Lavater." *Revue de littérature comparée* 1 (1921): 285–94.

Lattre, Alain de. *Le Réalisme selon Zola: archéologie d'une intelligence.* Paris: Presses Universitaires de France, 1975.

Laugaa, Maurice. "L'Effet *Fille aux yeux d'or.*" *Littérature* 20 (December 1975): 62–80.

Lavater, Johann Caspar. *Essai sur la physiognomonie, destiné à faire connoître l'homme et à le faire aimer.* 4 vols. Translated by Mme. E. de la Fîte, H. Renfner, and Mme. Caillard. The Hague, 1781–1803.

Lavater, Johann Caspar. *Essays on Physiognomy; for the promotion of the knowledge and the love of mankind.* 2d ed. Translated by Thomas Holcroft. London, 1804.

Lavater, Johann Caspar. *Physiognomische Fragmente, zur Beförderung der Menschenerkenntnis und Menschenliebe.* 4 vols. Leipzig, 1775.

Le Lavater des dames, ou l'art de connaître les femmes sur leur physionomie. Paris: Saintin, 1812.

Le Lavater portatif. Paris: Saintin, 1815.

Le Lavater portatif. 8th ed. Brussels: A. Wahlen, 1826.

Le Brun, Charles. "Conférence sur l'expression des passions," *Nouvelle Revue de psychanalyse* 21 (Spring 1980): 95–121.

Le Huenen, Roland, and Paul Peron, eds. *Le Roman de Balzac: recherches critiques, méthodes, lectures.* Montreal: Didier, 1980.

Lloyd, Rosemary. "Rereading *Mademoiselle de Maupin.*" *Orbis Litterarum: International Review of Literary Studies* 41, 1 (1986): 19–32.

Lloyd, Rosemary. "Speculum Amantis, Speculum Artis: The Seduction of *Mademoiselle de Maupin.*" *Nineteenth-Century French Studies* 15, 1–2 (Fall–Winter 1986–87): 77–86.

Locke, John. *An Essay concerning Human Understanding*. Edited by A. Fraser. New York: Dover Publications, 1959.

Lucas, Prosper. *Traité philosophique et physiologique de l'hérédité naturelle dans les états de santé et de maladie du système nerveux* . . . Paris: J. B. Baillière, 1850.

Lukács, Georges. *Balzac et le réalisme francais*. Paris: F. Maspero, 1967.

Malinas, Yves. *Zola et les hérédités imaginaires*. Paris: Expansion Scientifique Française, 1985.

Marivaux, Pierre Carlet de Chamblain de. *Journaux et œuvres diverses*. Edited by Frédéric Deloffre et Michel Gilot. Paris: Garnier, 1969.

Marivaux, Pierre Carlet de Chamblain de. *La Vie de Marianne*. Edited by Marcel Arland. Paris: Garnier-Flammarion, 1978.

Marshall, David. *The Surprising Effects of Sympathy*. Chicago: University of Chicago Press, 1988.

Matoré, Georges. *Le Vocabulaire de la prose littéraire de 1833 à 1845: Théophile Gautier et ses premières oeuvres en prose*. Geneva: Droz, 1951.

Matthews, J. H. *Les deux Zola: science et personnalité dans l'expression*. Geneva: Droz, 1957.

Michie, Helena. *The Flesh Made Word: Female Figures and Women's Bodies*. New York and Oxford: Oxford University Press, 1987.

Miller, Nancy K. *The Heroine's Text*. New York: Columbia University Press, 1980.

Mirabeau, Honoré Gabriel Riqueti, comte de. *Lettre à M. . . . sur M.M. Cagliostro et Lavater*. Berlin: François de la Garde, 1786.

Mitterand, Henri. *Zola et le naturalisme*. Paris: Presses Universitaires de France, 1986.

Molino, Jean. "Orgueil et sympathie à propos de Marivaux." *Dix-Huitième Siecle* 14 (1982): 337–55.

Morel, Bénédict-Auguste. *Traité des dégénérescences physiques, intellectuelles et morales de l'espèce humaine, et des causes qui produisent ces variétés maladives*. Paris: J. B. Baillière, 1857.

Mosse, George Lachmann. *Toward the Final Solution: A History of European Racism*. New York: H. Fertig, 1978.

Muhlemann, Suzanne. *Ombres et lumières dans l'œuvre de Marivaux*. Bern: Editions Herbert Lang et Cie., 1970.

Mylne, Vivienne. *The Eighteenth-Century French Novel: Techniques of Illusion*. Manchester: Manchester University Press, 1965.

Naudin-Patriat, Françoise. *Ténèbres et lumière de l'argent: la représentation de l'ordre social dans Les Rougon-Macquart*. Dijon: Université de Dijon, 1981.

Nelson, Brian. *Emile Zola: A Selective Analytical Bibliography*. London: Grant and Cutler, 1982.

Nelson, Brian. *Zola and the Bourgeoisie: A Study of Themes and Techniques in Les Rougon-Macquart*. Totowa, N.J.: Barnes and Noble, 1983.

Nykrog, Per. *La Pensée de Balzac dans La Comédie humaine*. Paris: Klincksieck, 1973.

P. . . . , M. *Le Lavater historique des femmes célèbres des temps anciens et modernes*. Paris: F. Louis, 1811.

Pagès, Alain. "En partant de la théorie du roman expérimental." *Cahiers naturalistes* 47 (1974): 70–87.

Pagès, Alain. "Rouge, jaune, vert, bleu: étude du système des couleurs dans *Nana*." *Cahiers naturalistes* 49 (1975): 125–35.

Palache, John Garber. *Gautier and the Romantics*. New York: Viking Press, 1926.

Parent-Duchâtelet, A. J. B. *De la prostitution dans la ville de Paris, considérée sous le rapport de l'hygiène publique, de la morale et de l'administration*. 2 vols. Paris, 1857.

Paris au XIXe siècle: aspects d'un mythe littéraire. Lyon: Presses Universitaires de Lyon, 1984.

Paulson, William R. "La Démarche balzacienne, ou le livre sur rien." *Romanic Review* 75, 3 (May 1984): 294–301.

Perrot, Philippe. *Le Corps féminin: le travail des apparences XVIIe–XIXe siècle*. Paris: Seuil, 1984.

Petrey, Sandy. "Obscenity and Revolution." *Diacritics* 3, 3 (Fall 1973): 22–26.

Pickup, Ian. "An Aspect of Portraiture in the Novels of Balzac: Psycho-Physiological Bridges." In *Literature and Society: Studies in Nineteenth and Twentieth Century French Literature Presented to R. J. North*, edited by C. A. Burns. Birmingham, England: Goodman for University of Birmingham, 1980.

Porta, Giovanni Batista della. *Della fisonomia dell'huomo*. Venice: Presso C. Tomasini, 1644.

Poulet, Georges. *Etudes sur le temps humain*. Paris: Plon, 1949.

Poulet, Georges. *Trois Essais de mythologie romantique*. Paris: José Corti, 1966.

Reboul, Jeanne. "Balzac et la vestiognomie." *Revue d'histoire littéraire de la France* 50, 2 (Apr.–June 1950): 210–33.

Renson, Jean. *Les Dénominations du visage en français et dans les autres langues romanes*. Paris: Les Belles Lettres, 1962.

Rey, Alain, ed. *Théories du signe et du sens*. 2 vols. Paris: Klincksieck, 1973.

Richard, Jean-Pierre. *Etudes sur le romantisme*. Paris: Seuil, 1971.

Richer, Jean. *Etudes et recherches sur Théophile Gautier, prosateur*. Paris: A. G. Nizet, 1981.

Roche, Daniel. *La Culture des apparences: une histoire du vêtement XVIIe–XVIIIe siècle*. Paris: Fayard, 1989.

Rogers, Samuel. *Balzac and the Novel*. Madison: University of Wisconsin Press, 1953.

Rosbottom, Ronald. "Marivaux and the Significance of Naissance." In *Jean-Jacques Rousseau et son Temps*, Edited by Michel Launay. Paris: A. G. Nizet, 1969.

Rosbottom, Ronald. *Marivaux's Novels: Theme and Function in Early Eighteenth-Century Narrative*. Rutherford, N.J.: Fairleigh Dickinson University Press, 1974.

Ross, Stephanie. "Painting the Passions: Charles Le Brun's *Conférence sur l'expression*." *Journal of the History of Ideas* 45, 1 (Jan.–March 1984): 25–47.

Rossiter, Andrew. "Le Sens de l'amour dans *Mademoiselle de Maupin* de Gautier." In *Hommages à Jacques Petit*, edited by Michel Maliclet. Paris: Les Belles Lettres, 1985.

Rousseau, Jean-Jacques. *Essai sur l'origine des langues*. Paris: Bibliothèque du Graphe, 1970.

Rousset, Jean. *Forme et signification*. Paris: José Corti, 1966.

Sadoff, Janet. "Ambivalence, Ambiguity and Androgyny in Théophile Gautier's *Mademoiselle de Maupin*." Ph.D. Diss., Harvard University, 1987.

Savalle, Joseph. *Travestis, métamorphoses, dédoublements: essai sur l'oeuvre romanesque de Théophile Gautier*. Paris: Librairie Minard, 1981.

Schaad, Harold. *Le Thème de l'être et du paraître dans l'œuvre de Marivaux*. Zurich: J. Druck, 1969.

Schapira, Marie-Claude. "L'Ecriture de Narcisse: fiction et reflexivité." *Bulletin de la société Théophile Gautier* 6 (1984): 109–23.

Schapira, Marie-Claude. *Le Regard de Narcisse: romans et nouvelles de Théophile Gautier*. Lyon: Presses Universitaires de Lyon; Paris: Editions du CNRS, 1984.

Schnack, Arne. "Surface et profondeur dans *Mademoiselle de Maupin*." *Orbis Literrarum: International Review of Literary Studies* 36, 1 (1981): 28–36.

Schor, Naomi. "Mother's Day: Zola's Women." *Diacritics* 5, 4 (Winter 1975): 11–17.

Schor, Naomi. "Le Sourire du sphinx: Zola et l'énigme de la féminité." *Romantisme* 13–14 (1976): 183–95.

Schor, Naomi. *Zola's Crowds*. Baltimore: Johns Hopkins University Press, 1978.

Serper, Arie. "Quelques Réflexions sur les personnages de Balzac et Zola." *Cahiers naturalistes* 56 (1982): 37–45.

Shortland, Michael. "The Power of a Thousand Eyes: Johann Caspar Lavater's Science of Physiognomical Perception." *Criticism* 28, 4 (Fall 1986): 379–408.

Shortland, Michael. "Secret Sins and Unnatural Follies." *Literary Review* (July 1982): 31–33.

Shortland, Michael. "Skin Deep: Barthes, Lavater and the Legible Body." *Economy and Society* 14 (1985): 273–314.

Slott, Katherine. "Narrative Tension in the Representation of Women in Zola's *L'Assommoir* and *Nana*." *L'Esprit créateur* 25, 4 (Winter 1985): 93–104.

Smith, Albert Brewster. *Ideal and Reality in the Fictional Narratives of Théophile Gautier*. Gainesville: University of Florida Press, 1969.

Spencer, Samia. "La Femme dans l'œuvre romanesque de Marivaux." *Language Quarterly* 15 (1976): 9–14.

Spitzer, Leo. "A Propos de *La Vie de Marianne*, lettre à M. Georges Poulet." *Romanic Review* 44 (1953): 102–26.

Spoelberch de Lovenjoul, Charles, vicomte de. *Histoire des oeuvres de Théophile Gautier*. Paris: G. Charpentier et Cie., 1887.

Stafford, Barbara Maria. *Body Criticism: Imaging the Unseen in Enlightenment Art and Medicine*. Cambridge, MA and London: MIT Press, 1991.

Stewart, Philip. *Half-Told Tales: Dilemmas of Meaning in Three French Novels*. Chapel Hill: North Carolina Studies in the Romance Languages and Literatures, 1987.

Stewart, Philip. *Imitation and Illusion in the French Memoir-Novel, 1700–1750: The Art of Make-Believe.* New Haven: Yale University Press, 1969.

Stewart, Philip. *Le Masque et la parole: le langage de l'amour au XVIIIe siècle.* Paris: José Corti, 1973.

Stewart, Philip. "Marianne's Snapshot Album: Instances of Dramatic Stasis in Narrative." *Modern Language Studies* 15, 4 (Fall 1985): 281–88.

Suffel, Jacques. "L'Odorat d'Emile Zola." *Aesculape* 33 (1952): 204–7.

Therrien, Madeleine. "La Problématique de la féminité dans *La Vie de Marianne.*" *Stanford French Review* 11, 1 (Spring 1987): 51–61.

Thomas, Ruth P. "'. . . et je puis dire que je suis mon ouvrage': Female Survivors in the Eighteenth-Century French Novel." *French Review* 60, 1 (October 1986): 7–19.

Thomson, Clive R. "Discours littéraire et discours idéologique: l'étude génétique des romans de Zola." *Cahiers naturalistes* 50 (1976): 202–12.

Tytler, Graeme. *Physiognomy in the European Novel: Faces and Fortunes.* Princeton: Princeton University Press, 1982.

Van Den Abbeele, Georges. "La Vérité en négligé: le 'Nouveau Monde' de Marivaux." *French Studies* 40, 1 (1986): 287–303.

Vannier, Bernard. *L'Inscription du corps: pour une sémiotique du portrait balzacien.* Paris: Klincksieck, 1972.

Van Rossum-Guyon, Françoise, and Michel van Brederode, eds. *Balzac et les parents pauvres.* Paris: SEDES, 1981.

Warren, Jill. "Zola's View of Prostitution in *Nana.*" In *The Image of the Prostitute in Modern Literature,* edited by Pierre Horn and Mary Beth Pringle. New York: Ungar, 1984. 29–41.

Wechsler, Judith. *A Human Comedy: Physiognomy and Caricature in Nineteenth Century Paris.* Chicago: University of Chicago Press, 1982.

Weil, Kari. *Androgyny and the Denial of Difference.* Charlottesville: U. of Virginia Press, 1992.

Weil, Kari. "Romantic Androgyny and its Discontents: The Case of *Mademoiselle de Maupin.*" *Romanic Review* 78, 3 (May 1987): 348–58.

Weil, Kari. "Veiling Desire: The Aesthetics of Androgyny in Balzac and Gautier." Ph.D. Diss., Princeton University, 1985.

Yale French Studies 42, "Zola" (1969).

Ysabeau, V. F. A. *Lavater et Gall: physionomie et phrénologie rendues intelligibles pour tout le monde.* Paris: Garnier Frères, 1862.

Zola, Emile. *L'Assommoir.* Paris: Le Livre de Poche, 1978.

Zola, Emile. *Carnets d'enquêtes: une ethnographie inédite de la France.* Edited by Henri Mitterand. Paris: Plon, 1986.

Zola, Emile. *Nana.* In vol. 2 of *Les Rougon-Macquart: histoire naturelle et sociale d'une famille sous le second empire,* edited by Henri Mitterand. Paris: Gallimard, 1961.

Zola, Emile. *Nana.* Edited by Roger Ripoll. Paris: Garnier-Flammarion, 1968.

Zola, Emile. *Le Roman expérimental.* Paris: Garnier-Flammarion, 1971.

Index